Niger Delta

A Path To Prosperity

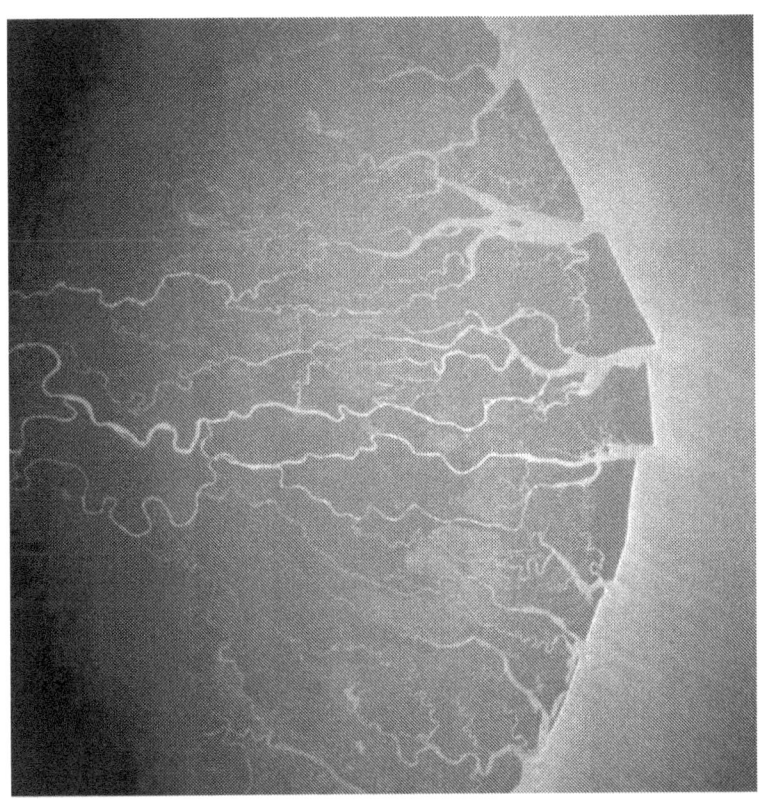

by
Tosan Alfred Rewane

authorHOUSE®

AuthorHouse™ UK Ltd.
500 Avebury Boulevard
Central Milton Keynes, MK9 2BE
www.authorhouse.co.uk
Phone: 08001974150

First published by AuthorHouse 9/4/2007

ISBN: 978-1-4343-3169-4 (sc)

Printed in the United States of America
Bloomington, Indiana

This book is printed on acid-free paper.

Acknowledgements

I wish to acknowledge the inspirational support of several friends and colleagues who have read the manuscript at various stages of its development. I thank them all for their insightful contributions and criticisms. David Hudson of Balliol College Oxford and a specialist on African affairs, in particular, has given formidable support in his creative and analytical review of the manuscript.

Special thanks go to Professor Winston Bellgam of the University of Port Harcourt, Rivers State for allowing the use of his research facilities in addition to his comments on the manuscript. Last but not the least my thanks go out to the research team of the *New Vision Foundation*, a new socio-economic think-tank for their tireless efforts in making this book a reality

Contents

The New Vision Foundation (NVF)

The NVF is a new policy research/advocacy enterprise. The Foundation's focus is the creation of an environment conducive to positive economic change in Nigeria, by promoting strategies and policies for strengthening Nigeria's economic performance. NVF creates policy platforms from which economic development can be built, in five strategic and inter-related areas:

- Wealth creation and competitiveness
- Poverty alleviation and economic inclusion
- Institutional structuring and innovation
- Sustainable development
- Social policy prioritisation

NVF recognizes the need for close public-private partnerships in formulating catalytic strategies for economic growth in Nigeria in particular and Africa in general. Such collaborations are especially important in Nigeria both because investors have often lacked confidence in the overall business environment and private sector initiative, energy and finance are critical to positive growth trajectories. To achieve higher growth rates, foreign investors perceptions of relatively high risks need to be addressed and overcome. Consequently the approach to transformation must reduce perceptions of risk, if it is to support the attainment of higher economic growth rates.

Prologue

Strategic Interest In the Niger Delta.

WASHINGTON, March 27, 2004 (UPI) -- President Bush is slated to meet with Nigerian President Olusegun Obasanjo in Washington Wednesday to discuss regional and energy security in West Africa, amid attacks on oil supply lines that U.S. officials consider strategic to U.S. "national interest."

Militants claiming to fight on behalf of a disenfranchised Niger Delta ethic minority released three foreign oil worker hostages Monday, but vowed to continue their three-month sabotage campaign that has cut Nigeria's daily oil exports of 2.5 million barrels by a quarter and contributed to international oil prices jumping to over $60 per barrel.

Although Nigeria's pipelines have long been vulnerable, the stakes have been raised as the United States increasingly depends on West African oil.

A surging energy demand in Asia and volatile climates in the Mideast and Latin America first prompted the Bush administration to call West African oil a "strategic national interest" in 2002 -- a label that freights the use of force to secure and defend such interests if necessary.

White House Press Secretary Scott McClellan reiterated Nigeria's role as a "strategic partner" in a Friday statement ahead of bilateral talks.

Nigeria stands as the fifth-largest supplier of oil to the United States, with 35.9 billion barrels of proven reserves. The shorter distances American tankers must travel to pick up crude that is both lighter and "sweeter" than the Middle-east variety has moved U.S. energy officials to stake the Gulf of Guinea will provide a quarter of U.S. crude by 2010.

This would place the region in front of Saudi Arabia as a leading oil supplier. Other major producers in the region include Equatorial Guinea, Angola, Gabon and Congo-Brazzaville.

The African Oil Policy Initiative Group, a Washington, D.C.-based lobby group with members from the oil industry and various arms of government, has recommended that Congress declare the Gulf of Guinea a "vital interest" to the U.S. and form a regional military sub-command, citing the "need to reshape a new U.S. national security policy for sub-Saharan Africa based on a West African economic engine driven by large petroleum revenues."

By providing the U.S. and other markets with a steady and secure flow of high-quality, reasonably priced African crude, dependence on hostile or unstable suppliers in other parts of the globe would diminish," the group concluded at a recent conference. But it remains to be seen whether a volatile patchwork of some 250 ethnic groups hastily pulled together more than a century ago by imperial Britain to feed London coffers can stay intact.

The 22 million inhabitants of the five delta states receive just 13 percent of oil revenues, while fishing waters are reportedly polluted without consequence by oil giants such as Shell and Chevron.

According to a 2004 World Bank report, 80 percent of Nigeria's oil wealth is usurped by just 1 percent of the population, the majority of which lives on less than $1 a day. Annual per capita income has dropped from $1,000 to $390 in the last 25 years, a phenomenon that has earned the country third-to-last ranking on Transparency International's 2005 corruption index.

Obasanjo the former president expanded press freedoms and spearheaded a campaign that led to the forgiveness of 60 percent of his nation's foreign debt. But pledges to purge the government of corruption have proved ineffective.

Oil theft, also known "bunkering," still exceeds 200,000 barrels a day, often thanks to the complicity of police who have not been paid in months.

Alternately, stepped up rebel attacks against oil facilities could move the military to depose the president and seize power.

May 30th, 2007 a new President of Nigeria, Umaru Musa Yar'adua is sworn in as the third civilian president of Nigeria, with a pledge, among other priorities to pay particular attention to the peoples of the Niger Delta in the overall interests of the country assuring justice and equity.

The Niger Delta is a vast flood plain situated in the coastal parts of Southern Nigeria and comprises nine States, namely: Rivers, Cross Rivers, Akwa Ibom, Bayelsa, Delta, Edo, Imo and Ondo States. It is the third largest wetland in the world; its rich flora and fauna has sustained life for generations.

As stated above the Niger Delta basin of Nigeria is the 5th largest delta basin in the world, and the second largest in Africa after the Nile basin. It is also one of the most petroliferous basins in the world. The Niger Delta has approximately 75,000 sq. km of sedimentary basin.

The region has a high quality hydrocarbon resource base. The oil and gas reserve base is strong with significant upside potential. The estimate of recoverable oil is about 35 billion barrels of high grade, light sweet crude oil. Over 57 multinational and private Nigerian firms have been licensed to explore and produce oil in the region from a total of 164 blocks. More blocks are soon to be awarded. Nigeria's average daily production is 2.4 million barrels per day, with installed production capacity of 3 million barrels per day.

The strategic importance of the Niger Delta region to the economic well being of Nigeria (and by extension, the whole sub-region and the industrialized economies) is thus very clear, hence the considerable attention being shown by local and international interests. It is this region, through it's hydrocarbon reserves that accounts for 90% of

Nigeria's economic wealth. Despite these vast natural resources, the Niger Delta has suffered great neglect that has brought about extreme poverty of the people and subsequent degradation of the environment.

As student at Cambridge University, England, I had often reflected on the paradox that is the Niger Delta; here is a region with vast natural resources, but remains one of the least developed regions of the world.

The overall thrust of any sustainable development strategy for the region should be to a) raise the standard of living of its citizens above the poverty line b) achieve the Millennium Development Goals (MDGs) and c) progress beyond these to enhance the technological and industrial capacity of the region. It should set out the specific policies and interventions that should be adopted to achieve these aims.

The strategy should also seek to create favourable conditions for greater productivity and economic enterprise in both rural and urban areas and by all sections of the population. It should also address specifically, the enabling conditions for enterprise, efficient agriculture and industrialization of the region utilizing the abundant natural resources in the region, and create the economic wealth needed to achieve the higher standard of living, while preserving the eco-system for long-term sustainability. The region faces significant environmental challenges, especially challenges resulting from gas flaring and venting. The Niger Delta region of Nigeria flares and vents gas equivalent to half its power consumption.

Infrastructure development, by and large, has been the main concern of government and its agencies, with varying degrees of success depending on the level of socio-political stability among the various host communities.

Niger Delta:
Rivers, States, & Vegetation

The remaining upstream potential of the Niger Delta is considerable, in the existing mature areas, in the exciting deepwater offshore Niger Delta and in shallow offshore swamp areas. The Oil & Gas sector has maintained a steady growth in reserves, buoyed by major reserves addition from deepwater acreages in recent years. Over 7billion barrels have been discovered in the deepwater acreages alone since 1996 and there is significant potential for more.

Production capacity has also grown steadily at an average annual rate of about 4% from about 1.25 million barrels per day to about 3 million barrels per day. In essence, the sector has grown production capacity at a rate higher than global average. Between 2005 and 2006, the oil sector added over 750 million barrels per day additional production capacity to global supplies mainly from deepwater projects.

Global energy is witnessing a new dawn characterized by steadily rising and high commodity price levels.

Going forward, there is a robust portfolio of Joint Venture (JV) and Production Sharing Contracts (PSC) projects planned to grow capacity to about 4 million barrels per day by 2010.

However, there is the challenge of improving local content in upstream operations, which has weak domestic economic linkages and so makes sub-optimal impact on job creation and economic growth in the region. With 3 million barrels produced per day, and assuming costs of $6 per barrel, 50% *local content* could mean the infusion into the regional economy of $9 million per day or about $3.2 billion per annum.

The major operators are committed to ending routine gas flaring by the year 2010. Renovation of aging production facilities and environmental remediation of polluted areas will account for massive additional investment from the oil companies.

Exploration and production of hydrocarbons in today's world is a matter of partnership. Partnership between the operators and the local communities, and partnership between the operators and other stakeholders beyond the actual operating arena. Critical short and medium term challenges to the Niger Delta will thus centre on peace and security issue and other key factors for its economic prosperity

In the Niger Delta, we should be focussed on creating a future that includes renewable energy. Our Nations growing dependence on fossil fuels and the impact of our consumption have never been more apparent than they are today. There is an alternative way-renewable fuels.

In this era of the new economy and globalisation, how can the private sector contribute to the eradication of poverty and the improvement in the living standards of people in the Niger Delta region of Nigeria in the shortest time possible? The emphasis is on improving living standards in the shortest time feasible.

The Federal Government of Nigeria (FGN) has tried to respond to discontent in the region by setting up a Niger Delta Development Commission (NDDC) with the mandate to develop the oil producing areas and offer a lasting solution to the socio-economic difficulties in the region. The FGN again recently set up the Council for the Sustainable Development of the Coastal States to give special attention to the coastal States in the Niger Delta.

In my modest effort to contribute to the national debate on how to move this strategic region forward, this book prioritises and examines the key factors that can, in my opinion put the Niger Delta region on the path to rapid and sustainable prosperity.

O N E

The New Economy & The Niger Delta

In this era of the new economy and globalization, how can the private sector contribute to the eradication of poverty and the improvement in the living standards of people in the Niger Delta in the shortest time possible? The emphasis is improving living standards in the shortest time feasible.

Modern information and communication technology (ICT) if properly applied, holds real prospects of meeting this challenge through the rapid integration of advanced and developing economies. In other words, technology driven globalization, can rapidly transform Niger Delta economies, by the rational engagement of the private sector in the mutual interest of global prosperity.

The term globalization in this book is defined as the dynamic process of global economic integration and consists of the following four key aspects:

- The removal of barriers to international trade. Companies increasingly, are engaged directly overseas as exporters or importers, and producers are exposed to competition of imports from the rest of the world.

- Increased interpenetration of markets by capital flows that can lift or wreck economies.

Internationalization of economic production. Major multinational corporations carry out an increasing proportion of international production. Although their headquarters are typically located in the industrialized countries, their production sites or areas of operation can be in any pert of the world.

Production itself is a highly logistical process that involves bringing components from one place, shipping them to another, dividing up what economists or business consultants call 'the value chain' and then farming out the individual parts of the production process to regions of the world where there is comparative advantage.

In one region, one handles logistical functions; in another region one takes advantage of low wages or exchange rate disparity; in yet another, of particular natural resources; and so forth. The importance of location for economic success has been enhanced by globalization of production process and its attendant benefits to the local citizenry.

Increased harmonization and standardization of economic policies, legislation and structures. Not only are countries becoming integrated into a web of production, a network of capital flows and an international market for goods and services, but also they engage in activities increasingly in a common structure of national and international institutions including environmental norms. The World Trade Organization (WTO) for trading norms, the IMF for macro-economic norms, the World Bank Group on norms for poverty eradication, norms on the enforcement of intellectual property etc. Patent law is now being "harmonized" internationally for the first time.

T W O

The ICT Imperitive

Modern information and communication technology (ICT), if properly applied, holds real prospects of meeting this challenge through the rapid integration of advanced and developing economies. In other words, technology driven globalisation, can rapidly transform the Niger Delta region economy, by the rational engagement of the private sector in the mutual interest of global prosperity.

The Digital Divide

Technology has been the main driver of globalisation. The advances achieved in computing and telecommunications in the West offer enormous, indeed unprecedented, scope for raising living standards in the Niger Delta. New technologies promise not just big improvements in local efficiency, but also the further and potentially bigger gains that flow from an infinitely denser network of connections, electronic and otherwise, with the developed world. These gains are not just profits for western and local corporations, but productive employment and higher incomes for the poor.

Much has been said about the so-called "Digital Divide" and the numerous strategies for bridging it. In 2001, the G-8 created the Digital Opportunity Task force or DOT force to this end. The conclusions in the group's first report were simple and unambiguous: Information and Communication technology (ICT) can provide 'powerful tools both for addressing people's basic needs and enriching the lives of poor people and communities in unprecedented ways.

In this book, I wish to go further by using the term "bridging the prosperity divide" to describe the path to bringing the wider benefits of modern technology, driven by the global private sector, to the Niger Delta.

The future development of ICT particularly in the rural Niger Delta will be largely private sector driven. The building of rural telecommunications infrastructure in particular will require significant amounts of investment. The Nigerian Communications Commission (NCC) in charge of policy and regulatory issues would want to see rural ICT implemented along commercial lines. The challenge involves:

* Building alliances and partnerships with private sector
* Attracting the required investment
* Raising the required capital for maintenance of networks
* Continuous equipment investment

Although this section of book focuses on the Niger Delta and the telecommunication sector as the basic model, it has wider implications for the rest of Africa, especially the highly indebted poorest nations.

One of the highlights of the DOT force recommendations was that every country needs an e-strategy as well as an emphasis on competitive suppliers for a communications infrastructure. In order to understand the impact of technologies on economies, it is important to be able to measure not only their effects, but also the extent of their deployment.

According to Jeffrey Sachs, countries across the globe are in three categories: technological innovators, technological adopters and technologically backward. Sach's innovators comprise about 15% of the globe's population and provide virtually all of the world's technology innovations. The adopters accounting for 55% of the world's population are able to adopt these technologies successfully in production and consumption. The rest of the world's population is not connected to technology in any significant way.

The United Nations Development Programme (UNDP) has developed an index for this purpose- the Technology Assessment Index (TAI) and it can be used in the same way as other indices (a popular one of which is the human development index (HDI), also developed by the UNDP.

The specific purpose of the TAI however, is to show how well a country is creating and diffusing technology and building a human skills base, thus reflecting it ability to participate in the "network age"- this being the combined result of the technological revolution and globalisation that is integrating markets and bringing prosperity to people all over the world.

But why is Africa lagging far behind in this process?

TAI is made up of four dimensions of technological capacity that are important for reaping the benefits of the network age and which measure actual achievements:

- Creation of technology – measuring innovation in society through patents and royalties and license flows from abroad
- Diffusion of recent innovations – measuring penetration of the Internet and the extent of high tech exports as a ratio of all exports
- Diffusion of old innovations – measuring telephony penetration and electricity consumption
- Human skills- measuring mean years of schooling and students enrolled in sciences such as mathematics and engineering.

It should come as no surprise that on a TAI list of 72 countries; the bottom ten consisted mainly of African countries. Many African countries did not even make the list. The highest African country on the list was South Africa.

There is a direct correlation between TAI and other development indices such as HDI and gross domestic product (GDP) per capita. Thus, in order to improve the overall development (and thus prosperity) of the continent, we have to improve TAI.

5

It is acknowledged that technology is a critical input into the global system of development. But it cannot be effective in isolation. There are a number of vital elements needed to ensure that technology is used to its fullest advantage. The first is that of appropriate government policy. It is all very well having access to the latest technologies, but if government policies result in the delayed or even worse, non-deployment of such technologies, then it becomes a waste of time

The situation in India today is vastly different. As it is the process of accelerating the development of new technologies, privatizing state-owned firms, dropping protection of many of its uncompetitive firms and opening its economy to increased foreign investment. India has opted to open dramatically the telecommunications sector, largely because keeping restrictions in place would halt India's e-commerce and information technology revolutions.

Thanks to this strategic decision, India is attracting new foreign investment and reversing the brain drain that has seen the best and brightest of its citizens leave the country due to lack of opportunities. Today, India is increasing educational opportunities for creating the kind of environment that will allow its people to build the country into the economic force it should be.

The regulation of ICT is heavily influenced by the changing role of government and business imperatives. Policymakers are often confronted by widely divergent cultural and economic values. For example, the computer industry is freewheeling, entrepreneurial, global and largely unregulated. The telecom sector is mostly nationalistic, monolithic and highly regulated.

Policymakers need to create new policies encouraging convergent services that combine:

- Computer services-that are unregulated and highly competitive
- Telecommunications-that are governed by regulated notions of equity and access

- Content- that is governed by regulated notions of what is right and proper in society.

There are only two viable outcomes:

- A hyper-completive, individualistic and libertarian approach, such as exists in the Internet and software industry or
- A new social contact between government and private enterprise, creating long-term vision and cohesion.

Promoting Private Investment in the Niger Delta

How can development agencies and government in the Niger Delta complement the private sector as a means of accelerating development and prosperity for its people.

There are two fundamental points. First, the private sector may often be better at assuming performance and market risks than public agencies. Shifting those responsibilities and risks to the private sector may thus be desirable. As part of this shift in risk allocation, the exposure of domestic taxpayers in poor countries to the burden of public debt can be reduced.

The second, when shifting risks, it is often desirable to unbundle some of the products of the World Bank Group. For example, by unbundling the policy risk function from the traditional Bank loan, policy risks guarantees have been created in the Bank and MIGA that shift commercial risks to private parties. There may now be further options available to shift performance risks to private parties so as to improve the investment climate and service to the poor. Unbundling the subsidy component embedded in some World Bank Group financial products would allow targeting it better at its purposes or beneficiaries.

Traditional aid in some sectors and in some low-income income countries such as in Africa has at times been associated with disappointing performance by state-owned agencies that were funded. In addition, subsidies embedded in aid funds may not have reached the poor but benefited the better off. There are thus two main issues:

How to improve service to the poor and how to ensure that any subsidy judged necessary actually benefit low-income communities like in the Niger Delta.

The key to better service delivery is to shift the performance risk more effectively to service providers (the private sector) and away from taxpayers. This can be achieved wherever it is possible to disburse aid when service is delivered and not when inputs are constructed. Such schemes are in principle, feasible in most parts of infrastructure and some areas of health and education. Service standards should be responsive to beneficiaries by empowering them to choose. This includes community participation to set goals for collective goods and services.

Providers (for profit and/or not-for-profit) can compete for the right to provide service on the best possible terms. Funding can be obtained in the market when the provider is competent and cash flow is expected to be adequate, which is facilitated when aid is disbursed upon achievement of contractual service obligations. The IFC could help to develop the market for this.

Policy risk guarantees by MIGA or the Bank CAN help can help deal with the special risks in difficult environments. In many complicated policy environments, small-scale solutions that are within the managerial and financial reach of domestic providers may provide appropriate solutions. (Note that output-based funding could be achieved with grants or with loans. In the case of loans, ultimately, the local taxpayer end up paying, while with grants, the foreign taxpayer pays-both in the case of concessional loans such as IDA).

Normal competitive markets are out-put based schemes. Consumers pay when the product or service is delivered, not when the factory is built.

Typically improved service delivery systems by themselves will help the poor most, because access to service rather than subsidies are crucial for them.

Niger Delta: Population Projection: High

STATE	2005	2010	2015	2020
Abia	3,230,000	3,763,000	4,383,000	5,106,000
Akwa Ibom	3,343,000	3,895,000	4,537,000	5,285,000
Bayelsa	1,710,000	1,992,000	2,320,000	2,703,000
Cross River	2,736,000	3,187,000	3,712,000	4,325,000
Delta	3,594,000	4,186,000	4,877,000	5,681,000
Edo	3,018,000	3,516,000	4,096,000	4,871,000
Imo	3,342,000	3,894,000	4,535,000	5,283,000
Ondo	3,025,000	3,524,000	4,105,000	4,782,000
Rivers	4,858,000	5,659,000	6,592,000	7,679,000
TOTAL	**28,856,000**	**33,616,000**	**39,157,000**	**45,715,000**

Source: NDR Survey-based on National Population Commission Data

N-DELTA HUMAN POVERTY INDEX FOR 2005				
STATE	Probability at birth of not surviving to age 40	Adult illiteracy rate	Unweighed average	HPI -1
Abia	26	26	34	29.169
Akwa Ibom	27	28	35.5	30.649
Bayelsa	30	31	39	33.826
Cross River	26	28	33	29.3
Delta	20	18	27	22.355
Edo	22	18	28	23.399
Imo	25	29	32	28.949
Ondo	30	31	42.5	35.442
Rivers	24	24	30.5	26.53
N-DELTA	**55.556**	**25.889**	**33.4**	**28.847**

HUMAN DEVELOPMENT INDEX FOR THE N-DELTA 2005				
STATE	**Life Expectancy**	**Education Index**	**GDP Index**	**HD 1**
Abia	0.492	0.578	0.560	0.543
Akwa Ibom	0.506	0.683	0.540	0.576
Bayelsa	0.455	0.523	0.520	0.499
Cross River	0.556	0.630	0.565	0.584
Delta	0.587	0.636	0.621	0.615
Edo	0.579	0.602	0.600	0.594
Imo	0.503	0.546	0.591	0.547
Ondo	0.501	0.575	0.512	0.529
Rivers	0.563	0.590	0.620	0.591
N-DELTA	**0.527**	**0.596**	**0.570**	**0.564**

Foreign direct investment, a subset of the overall flow of private finance, has become an important factor in bridging the "prosperity divide". The flow of money into stock markets in by short-term equity investments is significant, but volatile. FDI tends to be much more stable, representing a long-term investment that benefits recipient countries by transferring state-of-the-art technology, stimulating and promoting growth. In some ways, how much FDI a country attracts is an excellent indicator and hope for a better, more prosperous future.

Developing regions such as the Niger Delta are becoming a more attractive destination point for FDI as indicated both in the percentage share and in the overall amount of inflows. There remains controversy over FDI flows however. Looking at FDI as a percentage of GDP seems more realistic. Poor countries and poor small countries in particular, have a limited absorptive capacity for investment. Although it may be desirable to increase private investment, it does not necessarily follow that it is practical to do so. It has been shown that many countries' flow of FDI. Is more linked as a percentage of their GDP.

Looking at the overall trends, especially for African regions, it is obvious that many of these regions could absorb more private investment than they are now receiving. Attracting FDI through multinational companies enables countries to gain access to advanced technologies and capital as well as modern management practices and linkages to the advanced developed markets. These linkages are especially important. Recent studies by UNCTAD have shown that traditionally, affiliates of multinational companies may account for as much as one third of the so-called inter-firm trade (i.e., trade within an MNC). This, of course, can be managed in away that increases the competitiveness of the host country by linking the FDI to domestic firms.

UNCTAD has traditionally focused on foreign direct investment in their analytical and technical assistance programs. Recently their have been discussion of redirecting some of the efforts on direct investment to cover the policies of not just the host (destination) countries, but also the home (origin) countries and expanding the program to cover foreign portfolio investment.

Some G77 countries believe that the uneven distribution of FDI among developing countries could be addressed through home country policies to influence direct investment decisions by MNCs to consider those countries receiving minimal FDI resource flows.

Unlike debt, FDI does not need to be serviced and cannot flee at short notice. As a result, providers of FDI themselves bear most of the financial risk attached to the investment, but this advantage comes at a price. Over the long term, FDI can be a more expensive form of finance than debt because the outflow of remitted profits usually gives this kind of investor a bigger return than a foreign bank or bondholder could expect to receive. All the same, FDI is not only less risky for the host country; it is also more productive, because of the technology and techniques that come with it. It is expensive for the host country's point of view, but good value. Other forms of foreign capital, and especially short-term bank debt, have led many a developing country into desperate trouble.

Locational determinants of FDI flows fall into 3 groups (the first being the most important):

* Economic determinants (natural resources, large domestic market, labour costs, etc)
* Policy frameworks of the host country
* Business facilitation measures

Multinationals were increasingly focusing on "created assets" of technology skills and infrastructure in deciding location of FDI

A multilateral investment framework could contribute to increasing FDI flows through greater transparency. Enabling environment alone is not always sufficient. The development objectives in International Investment Agreements (IIAs) must include ways of attracting FDI flows and increasing transparency as a way to enhance development. All these issues must be considered when drawing up or reviewing new investment policies by government.

The harmonisation of new policies on a regional basis is also important and critical. The International Telecommunications Union (ITU) through the creation of the Africa Telecom Regulators Network took a step in this director.

The public sector does not have the funding capacity for large-scale ICT investments. Government must thus make themselves more attractive to private investment through fiscal and other incentives, which should include tax holidays; a reduction in import duties on certain equipment; and the possible lifting of restrictions on the free movement of labour.

Tony Blair, to 'take a fresh look at Africa's past and present and the international community's role in its development path', launched the Commission for Africa (CFA) in February 2004.

It wants developed countries to double the amount of aid to Africa, massive debt forgiveness for the highly indebted African nations, as well as boosting trade on more favorable terms The CFA sees the private sector, especially multinational corporations as playing a key role in the new initiative. Its plan is to invest massive amounts of G8 and African money in public-private partnerships to build the infrastructure that will eventually turn Africa into a single free market economy, a major international trading partner and a new growth site for foreign investment.

As Haiko Alfeld, director of the Africa World Economic Forum illustrates, business is clearly thrilled by the outcome of Blair's Commission for Africa. The OECD Guidelines on Multinational Companies should provide 'clear enough guidance on what companies should do with regards to their host communities'. Their 'corporate social responsibility' (CSR) policies must have greater impact and their operations more transparent.

The "Business Contact Group" represents the first time a G8 president has formally sought ideas from the U.S. private sector to shape discussion at the G-8 Summit.'

In July 2004 the then Chancellor Gordon Brown (now Prime Minister) and Reuter's chairman, Niall Fitzgerald, set up a Business Contact Group explicitly to provide private sector input to the Commission for Africa. Stakeholders in the Niger Delta could exploit this window of opportunity.

THREE

Production & Exploration
in the Niger Delta

As of 2000, oil and gas exports accounted for more than 98% of export earnings and about 83% of federal government revenue, as well as generating more than 40% of its GDP. It also provides 95% of foreign exchange earnings, and about 65% of government budgetary revenues.

Nigeria's proven oil reserves are estimated by the US EIA (United States Energy Information Administration) at between 16 and 22 billion barrels, but other sources claim there could be as much as 35.3 billion barrels. Its reserves make Nigeria the tenth most petroleum-rich nation, and by the far the most affluent in Africa. Nearly all of the country's primary reserves are concentrated in and around the delta of the Niger River, but off-shore rigs are also prominent in the well-endowed coastal region.

Nigeria is one of the few major oil-producing nations still capable of increasing its oil output and unlike most of the other OPEC countries, Nigeria is not projected to exceed peak production until at least 2009. The reason for Nigeria's relative unproductivity is primarily OPEC regulations on production in order to regulate prices on the international market. More recently, production has been forced to a halt intermittently by the demands and actions of the Niger Delta's inhabitants who feel they are being exploited.

Nigeria has over 259 total oil fields and 2481 wells in operation according to the Ministry of Petroleum Resources. The most productive region of the nation is the coastal Niger Delta Basin in the Niger Delta or "South-south" region which encompasses 78 of the 259 oil fields.

Many of Nigeria's oil fields are small and scattered, and as of 1990, these small unproductive fields accounted for 62.1% of all Nigerian production. This contrasts with the sixteen largest fields which produced 37.9% of Nigeria's petroleum at that time. As a result of the numerous small fields, an extensive and well-developed pipeline network has been engineered to transport the crude. Also due to the lack of highly productive fields, money from the jointly operated (with the federal government) companies is constantly directed towards petroleum exploration and production.

Much of Nigeria's petroleum is classified as "light" or "sweet", meaning the oil is largely free of sulphur. Nigeria is the largest producer of sweet oil in OPEC. This sweet oil is similar in constitution to petroleum extracted from North Sea. This crude oil is known as "Bonny light". Names of other Nigerian crudes, all of which are named according to export terminal, are Qua Ibo, Escravos blend, Brass river, Forcados, and Pennington Anfan.

In terms of exportation, the U.S. remains Nigeria's largest customer for crude oil, accounting for 40% of the country's total oil exports; Nigeria provides about 10% of overall U.S. oil imports and ranks as the fifth-largest source for U.S. imported oil.

There are six petroleum exportation terminals in the country, Shell owns two, while Mobil, Chevron, Texaco, and Agip own one each. Shell also owns the Forcados Terminal, which is capable of storing 13 million barrels of crude oil in conjunction with the nearby Bonny Terminal. Mobil operates primarily out of the Qua Ibo Terminal in Akwa Ibom State, while Chevron owns the Escravos Terminal located in Delta State and has a storage capacity of 3.6 million barrels. Agip operates the Brass Terminal in Brass, a town 113 km southwest of Port Harcourt and has a storage capacity of 3,558 barrels. Texaco operates the Pennington Terminal.

Natural Gas

Natural gas reserves are well over 100 trillion ft^3 (2,800 km^3), the gas reserves are three times as substantial as the crude oil reserves. The biggest natural gas initiative is the Nigerian Liquified Natural Gas Company, which is operated jointly by several companies and the state. It began exploration and production in 1999. Chevron is also attempting to create the Escravos Gas Utilization project which will be capable of producing 160 million standard ft^3 of gas per day.

There is also a gas pipeline, know as the West African Gas Pipeline, in the works but has encountered numerous setbacks. The pipeline would allow for transportation of natural gas to Benin, Ghana, Togo, and Cote d'Ivoire. The majority of Nigeria's natural gas is flared off and it is estimated that Nigeria loses 18.2 million USD daily from this.

Downstream

Nigeria's total petroleum refining capacity is 445,000 barrels per day (bpd), however, only 240,000 bpd was allotted during the 1990s. Subsequently crude oil production for refineries was reduced further to as little as 75,000 bpd during the regime of Sanni Abacha. There are four major oil refineries: the Warri Refinery and Petrochemical Plan which can process 125,000 barrels of crude per day, the New Port Harcourt Refinery which can produce 150,000 barrels per day (there is also an 'Old' Port Harcourt Refinery with negligible production), as well as the now defunct Kaduma Refinery. The Port Harcourt and Warri Refineries both operate at only 30% capacity.

The Port Harcourt and Kaduna refinery were recently acquired by a consortium of Nigerian investors under the Privatisation Programme.

It is estimated that demand and consumption of petroleum in Nigeria grows at a rate of 12.8% annually. However, petroleum products are unavailable to most Nigerians and are quite costly, because almost all of the oil extracted by the multinational oil companies is refined overseas, while only a limited quantity is supplied to Nigerians themselves.

Joint Venture Companies

All petroleum production and exploration is taken under the auspices of joint ventures between foreign multi-national corporations and the Nigerian federal government. This joint venture manifests itself as the Nigerian National Petroleum Corporation, a nationalized state corporation. All companies operating in Nigeria obey government operational rules and naming conventions (companies operating in Nigeria must legally be sub-entities of the main corporation, often incorporating "Nigeria" into its name). Joint ventures account for approximately 95 percent of all crude oil output, while local independent companies operating in marginal fields account for the remaining 5 percent.

Additionally, the Nigerian constitution states that all minerals, oil, and gas legally belong to the federal government. Six companies are operating in Nigeria and are listed with their countries of origin (most of the following is extracted from a Human Rights Watch report):

Royal Dutch Shell (British/Dutch)

Shell Petroleum Development Company of Nigeria Limited (SPDC), usually known simply as Shell Nigeria: A joint venture operated by Shell accounts for fifty percent of Nigerian's total oil production (899,000 barrels per day (bpd) in 1997) from more than eighty oil fields. The joint venture is composed of NNPC (55%), Shell (30%), TotaFinaElf (10%) and Agip (5%) and operates largely onshore on dry land or in the mangrove swamp in the Niger Delta. "The company has more than 100 producing oil fields, and a network of more than 6,000 kilometres of pipelines, flowing through 87 flowstations. SPDC operates 2 coastal oil export terminals".

The Shell joint venture produces about 50 percent of Nigeria's total crude. Shell Nigeria owns concessions on four companies, they are: Shell Petroleum Development Company (SPDC), Shell Nigeria Exploration and Production Company (SNEPCO), Shell Nigeria Gas (SNG), Shell Nigeria Oil Products (SNOP), as well as holding a major stake in Nigeria Liquified Natural Gas (NLNG). Shell formerly operated alongside British Petroleum as Shell-BP, but BP has since sold all of its Nigerian concessions. Most of Shell's operations in Nigeria are conducted through the Shell Petroleum Development Company (SPDC).

Chevron (American)

Chevron Nigeria Limited (CNL): A joint venture between NNPC (60%) and Chevron (40%) has in the past been the second largest producer (approximately 400,000 bpd), with fields located in the Warri region west of the Niger river and offshore in shallow water. It is reported to aim to increase production to 600,000 bpd.

Exxon-Mobil (American)

Mobil Producing Nigeria Unlimited (MPNU): A joint venture between the NNPC (60%) and Exxon-Mobil (40%) operates in shallow water off Akwa Ibom state in the southeastern delta and averaged production of 632,000 bpd in 1997, making it the second largest producer, as against 543,000 pbd in 1996.

Mobil also holds a 50% interest in a Production Sharing Contract for a deep water block further offshore, and is reported to plan to increase output to 900,000 bpd by 2000. Oil industry sources indicate that Mobil is likely to overtake Shell as the largest producer in Nigeria within the next five years, if current trends continue, mainly due to its offshore base allowing it refuge from the strife Shell has experienced onshore. It is headquartered in Eket and operates in Nigeria under the subsidiary of Mobil Producing Nigeria (MPN).

Agip (Italian)

Nigerian Agip Oil Company Limited (NAOC): A joint venture operated by Agip and owned by the NNPC (60%), Agip (20%) and ConocoPhillips (20%) produces 150,000 bpd mostly from small onshore fields.

Total (French)

Total Petroleum Nigeria Limited (TPNL): A joint venture between NNPC (60%) and Elf (now Total) produced approximately 125,000 bpd during 1997, both on and offshore. Elf and Mobil are in dispute over operational control of an offshore field with a production capacity of 90,000 bpd.

Texaco (now merged with Chevron)

NNPC Texaco-Chevron Joint Venture (formerly Texaco Overseas Petroleum Company of Nigeria Unlimited): A joint venture operated by Texaco and owned by NNPC (60%), Texaco (20%) and Chevron (20%) currently produces about 60,000 bpd from five offshore fields.

Imapct of Oil Industry on the Environment

Map of vegetation in Nigeria

As mentioned earlier in the book, the Niger Delta is comprised of 70,000 km² of wetlands formed primarily by sediment deposition. Home to 20 million people and 40 different ethnic groups, this floodplain makes up 7.5% of Nigeria's total land mass. It is the largest wetland and maintains the third-largest drainage area in Africa. The Delta's environment can be broken down into four ecological zones: coastal barrier islands, mangrove swamp forests, freshwater swamps, and lowland rainforests.

This incredibly well-endowed ecosystem, which contains one of the highest concentrations of biodiversity on the planet, in addition to supporting the abundant flora and fauna, arable terrain that can sustain a wide variety of crops, economic trees, and more species of freshwater fish than any ecosystem in West Africa. The region could experience a loss of 40% of its inhabitable terrain in the next thirty years because of extensive dam construction in the region. The carelessness of the oil industry has also precipitated this situation, which can perhaps be best encapsulated by a 1983 report issued by the NNPC in 1983, long perform popular unrest surfaced:

Oil spills in the Niger Delta occur due to a number of causes, they include: corrosion of pipelines and tankers (accounts for 50% of all spills), sabotage (28%), and oil production operations (21%, with 1% of the spills being accounted for by inadequate or non-functional production equipment.

The largest contributor to the oil spill total, corrosion of pipes and tanks, is the rupturing or leaking of production infrastructures that are described as, "very old and lack regular inspection and maintenance". A reason that corrosion accounts for such a high percentage of all spills is that as a result of the small size of the oilfields in the Niger Delta, there is an extensive network of pipelines between the fields, as well as numerous small networks of flowlines—the narrow diameter pipes that carry oil from wellheads to flowstations—allowing many opportunities for leaks. In onshore areas, most pipelines and flowlines are laid above ground. Pipelines, which have an estimate life span of about fifteen years, are old and susceptible to corrosion. Many of the pipelines are as old as twenty to twenty-five years. Even Shell admits that "most of the facilities were constructed between the 1960s and early 1980s to the then prevailing standards. SPDC [Shell Petroleum and Development Company] would not build them that way today." Shell operates the Bonny Terminal in Rivers State, which has reportedly been in operation for forty years without a maintenance overhaul; its original lifespan was supposed to be twenty five years.

Sabotage is performed primarily through what is known as "bunkering",whereby the saboteur attempts to tap the pipeline, and in the process of extraction sometimes the pipeline is damaged or destroyed. Oil extracted in this manner can often be sold for cash compensation.

The Nigerian National Petroleum Corporation places the quantity of oil jettisoned into the environment yearly at 2,300 cubic meters with an average of 300 individual spills annually. However, because this amount does not take into "minor" spills, the World Bank argues that the true quantity of oil spilled into the environment could be as much as ten times the officially claimed amount. Among the largest individual spills include the blowout of a Texaco offshore station which in 1980 dumped an estimated 400,000 barrels of crude into the Gulf of Guinea and Shell's Forcados Terminal tank failure which produced a spillage estimated at 580,000 barrels. One source projects that the total amount oil in barrels spilled between 1960 and 1997 is upwards of 100 million barrels.

Oil spillage has a major impact on the ecosystem into which it is released. Immense tracts of the mangrove forests, which are especially susceptible to oil (this is mainly because it is stored in the soil and re-released annually with inundation), have been destroyed. An estimated 5-10% of Nigerian mangrove ecosystems have been wiped out either by settlement or oil. The rainforest which previously occupied some 7,400 km² of land has disappeared as well.

Spills in populated areas often spread out over a wide area, taking out crops and aquacultures through contamination of the groundwater and soils.

Though the consumption of dissolved oxygen by bacteria feeding on the spilled hydrocarbons also contributes to the death of fishes. In agricultural communities, often a year's supply of food can be destroyed by only a minor leak, debilitating the farmers and their families who depend on the land for their livelihood. Drinking water is also frequently contaminated, and a sheen of oil is visible in many localized bodies of water. If the drinking water is contaminated, even if no immediate health effects are apparent, the numerous hydrocarbons and chemicals present in oil are highly carcinogenic. Although, people often do manifest sickness following consumption of polluted water.

Offshore spills, which are usually much greater in scale, contaminate coastal environments and cause a decline in local fishing production.

The decline in ecologic sustainability parallels the increase in oil production since operations began four decades ago. Furthermore, operating companies such as Shell have made public proposals for increasing production significantly in the near future which, because of the careless nature of oil operations in the Delta, will cause the environment to grow increasingly uninhabitable.

Natural Gas Flaring

Nigeria flares more natural gas associated with oil extraction than any other country on the planet, with estimates suggesting that of the 3.5 billion cubic feet of associated gas (AG) produced annually, 2.5 billion cubic feet, or about 70% is wasted via flaring. This equals about 25% of the UK's total natural gas consumption, and is the equivalent to 40% of the entire African continent's gas consumption in 2001. All statistical data associated with gas flaring is notoriously unreliable, but AG wasted during flaring is estimated to cost Nigeria US $2.5 billion on a yearly basis.

The reason for this practice, which is generally agreed world-wide to be wasteful both economically and environmentally, is that in order to maximize production of crude oil, the associated gas accompanying it is often burned off. Even though companies operating in Nigeria also harvest natural gas for commercial purposes, they prefer to extract natural gas from deposits where it is found in isolation, this isolated gas is known as non-associated gas. This occurs because it is costly to separate commercially viable associated gas from the oil. Therefore the AG found with oil is often burned off in order to increase crude production.

Historically, gas flaring began simultaneously with oil extraction in the 1960s by Shell-BP. Although, the British government subsequently acknowledged that the flaring was unacceptable, it was allowed to continue without any real efforts to change infrastructure and prevent the waste of the gas. This is in contrast to Britain's policies on gas flaring in their own territory, where gas flaring has been reduced to a minimum.

In fact, in western Europe 99% of associated gas is used or re-injected into the ground. Gas flaring is generally discouraged and condemned by the international community, as it contributes greatly to climate change. Which ironically can display its most devastating effects in developing countries like Nigeria, and particularly in the semi-arid Sahel regions of sub-Saharan Africa. The Niger Delta's low-lying plains are also quite vulnerable as they lie only a few meters above sea-level.

Gas flaring in Nigeria is also highly inefficient and releases large amounts methane, which has very high global warming potential. The methane is accompanied by the other major greenhouse gas, carbon dioxide, of which Nigeria was estimated to have emitted more than 34.38million tons in 2002, accounting for about 50% of all industrial emissions in the country and 30% of the total CO2 emissions. As flaring in the west has been minimized, in Nigeria it has grown proportionally with oil production. The volume of associated gas produced and therefore burnt off, is directly linked to the amount of oil produced. So even though the percentage of gas flared from 92% in 1981 has fallen to around 70%, the overall amount of flared gas has increased from 2.1 billion cubic feet to 2.5 billion cubic feet.

It seems that the international community, the Nigerian government, and the oil corporations are all in agreement that gas flaring has a negative impact and needs to be stopped. However, in reality, efforts at stemming gas flaring have been slow to be implemented. The practice of gas flaring as it has been allowed since oil production began under British, has become set in stone, and would be costly to overhaul to reduce flaring. As a result, little is done by oil companies. This is in spite of the fact that gas flaring in Nigeria has technically been illegal since 1984 under section 3 of the "Associated Gas Reinjection Act". However, none of the regulations stipulated by this document have ever been made public.

OPEC and Shell, the biggest flarer of natural gas in Nigeria, alike claim that only 50% of all associated gas are burnt off via flaring at present. However, this statistic is accepted by few. The World Bank reported in 2004 that, "Nigeria currently flares 75% of the gas it produces. While other sources make similar projections, between 70 and 75% is the generally accepted percentage of gas flared.

Gas flares can have potentially harmful effects on the health and livelihood of the communities in their vicinity, as they release a variety of poisonous chemicals. Just some of combustion by-products include nitrogen dioxides, sulphur dioxide, volatile organic compounds like benzene, toluene, xylene and hydrogen sulfide, as well as carcinogens like benzapyrene and dioxin.

Humans exposed to such substances can suffer from a variety of respiratory problems, which have been reported amongst many children in the Delta but have apparently gone uninvestigated. These chemicals can aggravate asthma, cause breathing difficulties and pain, as well as chronic bronchitis. Of particular note is that the chemical benzene, which is known to be emitted from gas flares in undocumented quantities, is well researched as being a causative agent for leukemia and other blood-related diseases. A study done by Climate Justice estimates that exposure to benzene would result in eight new cases of cancer yearly in Bayelsa State alone.

Often gas flares are located close to local communities, and regularly lack adequate fencing or protection for villagers who may risk nearing the tremendous heat of the flare in order to carry out their daily activities. Many of these communities claim that nearby flares cause acid rain which corrodes their homes and other local structures, many of which have metal roofing.

However, whether or not the flares contribute to acid rain is debatable, as some independent studies conducted have found that the sulphur dioxide and nitrous oxide content of most flares was insufficient to establish a link between flaring and acid rain. Other studies from USEIA (U.S. Energy Information Administration) report that gas flaring is, "major contributor to air pollution and acid rain".

Almost no vegetation can grow in the area directly surrounding the flare due to the tremendous heat it produces.

F O U R

Opportunities for Collaboration between Multinational Oil Companies and Local Indigenous Oil Companies

One of the ways Government the private sector (represented by the oil & gas companies) can show greater commitment to host communities in the Niger Delta directly affected by oil & gas exploration activities is to allow for greater participation of new entrants or existing operationally active local indigenous oil companies which should be persuaded to allocate not less than 20% of their share capital to the host communities that immediately surround the operating fields.

These companies can in turn enter into collaborative arrangements with the multinationals for the operation of *marginal fields*, other fields or a proportion of the oil wells that have been shut down due to high risks associated with youth militancy or conflicts of one nature or the other.

This will allow the multinationals to focus on deep and ultra-deep off-shore fields, while the presence of local indigenous companies representing a wide spread of community interests will help to reduce tension around on-shore production activities. The deep and ultra deep offshore offers significantly less risks in terms of militancy and conflicts.

The communities can be encouraged to organize themselves into co-operatives that will hold the share holding interest in the local oil companies with a minimum of one representative on the board of the local company.

Local indigenous companies have shown a clear ability to successfully operate fields that are not profitable to the majors. Why is this? They both receive the same price for the oil and gas they sell. The majors even have and advantage in getting their product to the market through the extensive transportation systems at lower costs than the local companies. Both are taxed at more or less equivalent rates and the major may be to more effectively use its deductions against a much broader income base. The annual returns will thus accrue directly to the communities that can be used for development purposes.

The main difference lies in the cost of operations. The cost for local companies comprises almost exclusively the actual directly related expense of operating their fields. The major, bears an additional and significant burden, namely overhead expenses from service groups within the organization that consume cash without generating a direct return.

Such overhead expense gets allocated among the various operating companies adding to their overall operating expense. Thus fields that may appear profitable relative to their direct operating expense can become marginal or even loss making under the burden of allocated overheads.

Constraints of Indigenous Oil Companies

The disadvantages the local companies have in the Niger Delta are the lack of transportation and terminal systems necessary to sell their product. This introduces the opportunity for collaboration through crude oil handling arrangements, and it is very possible for such arrangements to benefit both the shipper and the receiver of the crude oil.

From the financial and technical standpoint, to successfully make an oil well in the Niger Delta is only the first step. Getting the oil to the market is quite another. Looking beyond the well head, the producer whether a local or major, needs gathering systems, processing facilities, transportation networks and an export terminal. These are capital intensive.

To illustrate the point of getting oil from the wellhead to the market place lets suppose a local company or a multinational discovers a 15,000 barrels per day oil field which is five miles away from a suitable location for a terminal. The producer then installs a flow-station that is connected by an eight-inch pipeline to the terminal that is constructed. Aside from the cost to drill, complete and tie in the well, which by itself is several millions of dollars, the producer faces a capital investment of in the order of US$55million. To be certain, nobody is going to install a separate terminal for 15,000 barrels of oil per day. Just as it is also certain that to be called a producer, one needs access to transportation and terminal facilities.

Crude Oil Handling Arrangements

However multinationals operating several fields often have spare capacity at their terminals that could provide the opportunity to handle new production from other companies (particularly local companies) in the Niger Delta. Thus an equitable crude handling arrangement with a multinational with available capacity makes the project viable.

Joint Ventures

The best form of collaboration would be one that brings the local company and the multinational together before the first well is drilled, i.e. active partnering through joint ventures. A number of joint venturing schemes are possible. The most basic of these would be a case where the local company joins ranks with a multinational company, with working interest established prior to any exploration of development.

The parties share in exploration and development cost and receives the benefits of a successful project according to their respective working interest. Operatorship of the venture could be by one party exclusively or responsibilities could be shared. The multinational gets to add a potentially business venture to its portfolio. The local indigenous company gains from the experience and the technical expertise of the multinational. Both parties benefit by leveraging themselves into a profitable venture or reducing their exposure to loss depending on the outcome of the venture.

Farm-in/Farm-out Schemes

One means of establishing a joint venture is by farm-in/farm-out arrangements where the operator of an oil lease may elect for a variety of reasons to assign (farm-out) some portion of its interest to a local company. Farm-in/farm-out arrangements can vary widely, but typically involve the farming-in party agreeing to pay the drilling or completion of one or more wells in exchange for earning an interest in the lease. The farming out party normally retains either a royalty or working interest in the project.

Joint Facilities Development.

The type of joint venture that limits the extent of the partnership to crude handling but goes beyond a simple crude handling arrangement is joint facilities development. One could envision joint facilities between the local company of either a straddled field or nearby fields. Both would develop and operate their own fields, but would share in the installation of common processing and transportation facilities. The relative investment of each party would be according to the allocated share of the system capacity.

The obvious advantages also include elimination of redundancy in facilities and personnel, possible capture of economies of scale, and the spreading of capital risk. The local company would compensate their partner for an appropriate share of the technical work. But that compensation would be far less than if those same technical services were contracted to a commercial design and construction firm.

Unitised Development.

For straddled reservoirs, by far the best option is unitized development. The concept of unitized development is simple enough. A provisional boundary encompassing the straddled and contiguous reservoirs within a given field is determined and agreed. One of the partners is designated to be the operator of the unit on the basis of participation factors, proximity to existing transportation and terminal facilities and other pertinent logistical considerations. The unit operator goes on to develop and operate the field as a single entity and according to responsibilities spelled out in a comprehensive unitization and unit development agreement.

Experience As an Asset.

One might say that the multinationals are cooperating with the local companies by virtue of the fact that they are already operating in the Niger Delta. They have succeeded in creating a successful petroleum producing industry in the country that also includes a diverse and well-populated service sector that offers choice and competitive pricing. Local indigenous companies arriving newly into the scene can only benefit by finding an already established infrastructure.

What's more, the service companies working closely with their clients gain expertise (drilling, completion, simulation etc). That experience is a valuable asset that benefits their other clients, including locals and multinationals alike.

Capacity Building Of Indigenous Staff.

TRAINING NEEDS

SECTOR: ENERGY INFORMATION SYSTEM

Skills/training priorities 2	Training institutions 3	Assistance required to enhance capacity of institutions 4	Preferred partner organisation	Sources of funds
1. Energy sector analysis 2. Energy data analysis and dissemination 3. Energy planning issues.	Specialist Universities	Staff training Equipment Funding for project management and expert team including travel	International energy associations National energy departments WEC Religional economic organization	National energy departments Foreign donors ESMAP International energy organizations

Another derivable benefit from the multinationals is in the area of training Nigerian staff particularly in the technical ranks. Forty-seven years of operating in the Niger Delta by the multinational has given them great experience that is being transferred to indigenous employees. The saying goes ' we train and the local companies gain'.

Collaboration in the Downstream Sector.

Having focused mainly on the upstream side of the Nigerian petroleum industry, one can also touch on the downstream sector and opportunities for collaboration that will surely develop there. There is indeed great promise for a burgeoning petrochemical sector. The local oil companies will be keen to secure a share of the market, supplying this industry with raw materials-oil, natural gas, and gas liquids etc.

Some of the same schemes mentioned for collaboration in the upstream sector could also apply to other production commodities. As well, niches may evolve for local indigenous companies to become gas gatherers and processors, chemical feedstock and product processors and the like. This would open a complete new era of cooperation and active partnering between multinationals and local downstream companies.

What is the Message?

The recurring message in all this is that cooperation brings benefits or advantages to the parties (including the host communities) who are involved, either directly or indirectly. Not the least in this regard, is the government, its various agencies, the Niger Delta citizens and the country as a whole. The Government is likewise in a position through its involvement and regulations, to either encourage or hinder such ventures.

The Crucial Role of Government Agencies.

The NNPC and the DPR occupy unique positions and they have a vantage point perspective. The NNPC, itself an example of a fully integrated major oil company, is now and will continue to be involved in collaborative activities with various local operatives. This may be directly through leasing agreements or indirectly through activities of operators of its joint ventures with the multinationals.

Some of the opportunities for cooperation described above are necessarily preceded by a fair amount of planning, discussions and negotiations. The NNPC and DPR by virtue of having dealings with all of the companies operating in the Niger Delta, have a bird's eye view of potential areas for cooperation. Where it may not be obvious to the prospective parties that such a situation exists, the NNPC and the DPR should offer the suggestions that can get collaboration off to an early rather than late start.

Collaborative activities, as far as possible, should be entered into freely and be mutually beneficial and equitable for all parties involved. So long as the players act reasonably and fairly, these traits should be met. Neither the NNPC nor the DPR should imposition itself in a manner that causes an inequitable or detrimental situation to occur.

Most importantly, the NNPC and DPR should remove roadblocks to active partnership and joint venturing between the indigenous oil companies and the multinationals who already operate in the Niger Delta and who form the backbone of a diverse and flourishing oil industry in the country.

FIVE

The Need for an Environmentally-Friendly Clean Technologies Initiative

Every year oil producers flare and vent gas equivalent to the combined gas consumption of Central and South America. The Niger Delta region of Nigeria flares and vents gas equivalent to half its power consumption. To reduce this practice, governments need to develop and enforce an appropriate legal and regulatory framework. The FGN needs to reform the gas market, getting rid of subsidies for competing fuels and making room for private operators to develop gas infrastructure. And they need to avoid tax and royalty systems that discourage operators from using the gas associated with oil production

The connection between current energy use patterns and environmental degradation has been clearly documented, and such degradation will be severely accelerated in the future if new technological solutions and new priorities are not brought into the development planning process.

A different approach to energy that places greater emphasis on energy efficiency enhanced and increased commercial use of renewable forms of energy and the introduction of modern cleaner energy technologies in Nigeria is required.

What is thus proposed is a new initiative, the National Clean Technologies Initiative (NCTI) the goal of which will be to promote and facilitate, including working with international agencies to hold regular training workshops and seminars.

NCTI will when fully set up, act as the primary catalyst in building the country's sustainable development technology infrastructure, i.e. the establishment of clean fuel production units that benefit the economy, the environment and the health of Nigerians.

Tomorrow's generation of high-efficiency automobiles and trucks will need gasoline and diesel fuels that are environmentally superior to traditional fuels and can result in cleaner air and fuel efficiency (more miles-per-gallon). This new generation of Liquid and Solid Fuels will include the following:

- Ultra Clean (low-sulfur) Transportation Fuels
- Natural Gas – to – Liquids
- Solid Fuels and Feed-stocks
- Coal-Based Liquid Fuels and Chemicals
- Bio-fuels (Ethanol, E85, bio-diesel)

In summary the goal of the National Clean Technologies Initiative (NCTI) is to promote projects in the Niger Delta that commercialize current and proven research and development(R &D) to produce fuels and chemicals that will:

- Enable vehicles comply with current and future emission requirements.
- Be compatible with the existing liquid fuels infrastructure.
- Enable vehicle efficiencies to be significantly increased, with concomitantly reduced CO_2 emissions.
- Be obtainable from a fossil resource, alone or in combination with other hydrocarbon materials such as refinery wastes, municipal wastes, coal and biomass.
- Be cost competitive with current fuels

NCTI can, when fully set up, act as the primary catalyst in building the country's sustainable development technology infrastructure, i.e. the establishment of clean fuel production units that benefit the economy, the environment and the health of residents of the Niger Delta in particular and Nigerians in general.

New technologies have become available for utilizing the Niger Delta's abundant gas resources by converting the resource into high value fuels, oils and chemicals.

The basis of this proposal is to establish a co-ordinated and organized system of promoting and implementing joint venture projects involving these latest gas-processing technologies between those with the technology and potential business partners in the private and/or public sector (e.g. the Nigerian National Petroleum Corporation and related companies) and Niger Delta Stake-holders. These endeavors will thus including promoting public-private partnerships (PPPs).

Capacity Building:

The connection between current energy use patterns and environmental degradation has been clearly documented, and such degradation will be severely accelerated in the future if new technological solutions and new priorities are not brought into the development planning process.

A different approach to energy that places greater emphasis on energy efficiency enhanced and increased commercial use of renewable forms of energy and the introduction of modern cleaner energy technologies in Nigeria is required. This is what NCTI is out to promote and facilitate, including working with international agencies to hold regular training workshops and seminars.

Tomorrow's generation of high-efficiency automobiles and trucks will need gasoline and diesel fuels that are environmentally superior to traditional fuels and can result in cleaner air and fuel efficiency (more miles-per-gallon). This new generation of Liquid and Solid Fuels will include the following:

- Ultra Clean (low-sulfur) Transportation Fuels
- Natural Gas – to – Liquids

- Solid Fuels and Feed-stocks
- Coal-Based Liquid Fuels and Chemicals
- Bio-fuels (Ethanol, E85, bio-diesel)

In summary the goal of the proposed National Clean Technologies Initiative (NCTI) is to promote projects in Nigeria that commercialize current and proven research and development(R &D) to produce fuels and chemicals that will:

- Enable vehicles comply with current and future emission requirements.
- Be compatible with the existing liquid fuels infrastructure.
- Enable vehicle efficiencies to be significantly increased, with concomitantly reduced CO_2 emissions.
- Be obtainable from a fossil resource, alone or in combination with other hydrocarbon materials such as refinery wastes, municipal wastes, coal and biomass.
- Be cost competitive with current fuels.
- Etc

Developing a National Clean Technologies Strategy:

The proposed NCTI will encourage, assist and work with government, and all stakeholders in the Niger Delta to develop a National Clean Technologies Strategy that will cover the use of associated gas, waste and other hydrocarbons, unveiling bottlenecks and other barriers to full economic usage of market potential for clean, high efficiency, energy products as well as provide a package of new legal and fiscal incentives to encourage current and potential investors in clean energy projects.

The strategy will provide the appropriate pricing structures for both feedstock and end products and address the issue of adequacy of infrastructure provision (whether upstream, midstream or downstream) that are necessary for exploiting the abundant business opportunities in clean energy production.

The GGFR

The NCTI project is in line with and supports the objectives of the Global Gas Flaring Reduction (GGFR) Program being pursued by the Federal Government of Nigeria in association with the World Bank. The Nigerian Government has set the target of ending all flaring by 2008, and has endorsed the Voluntary Standard for Global Flaring and Venting Reduction developed by GGFR.

The GGFR, whose global office is in Washington, DC20433, USA is a forum of governments of oil-producing countries, state owned oil companies, international oil companies, as well as other key stakeholders, led by the World Bank Group, which supports the efforts of the petroleum sector Worldwide to progressively reduce emissions related to crude oil production.

The GGFR developed the Voluntary Standard for Global Flaring and Venting reduction after extensive consultations with the oil and gas industry and the governments of the countries where flaring occurs. (Details of the Standard can be found on http://www.worldbank.org/ggfr)

The implementation of the Standard is intended to reduce venting and flaring significantly within 5-10 years through:

- Providing a framework for governments, companies, and other key stakeholders to take complementary and supportive action;
- Encouraging an integrated approach including market and infrastructure development, commercialization, legal and fiscal regulations, and carbon credits, and
- Promoting stakeholder consultation and collaborative action to help reduce barriers to associated gas use in the country.

Part of the GGRC objective is to identify potential areas/project categories for associated gas use (with special attention to gas-to-power and gas-to-liquids).

The Clean Development Mechanism (CDM)

The NCTI project amply supports the carbon credit program under the CDM considering the reliability, cleanliness, economy and efficiency of the Clean Fuels as a substitute for petroleum-based fuels such as Kerosene that emits substantial GHG.

The increase in Nigeria's natural gas reserves particularly in areas of the Niger Delta remote from conventional gas markets is well known. Options for the development and uses of these remote gas fields are limited. Pipelining the gas to markets is possible but is capital intensive, and except in exceptional cases (like the W.A.G.P) limited to short distances.

The production of LNG has the problem of limited markets and the investment involved requires that market contracts be fixed prior to development of the project. The programs described in this proposal offer feasible alternatives for investment by the private sector standing alone or in joint venture with public sector agencies.

The vast natural gas reserves of the Niger Delta have been without market potential because they are simply too remote from the Nation's pipelines and urban markets. Advances in chemical conversion of natural gas suggest an added option of bringing associated natural gas to markets.

Using the chemical GTL technology, the gas is converted into liquids that are more easily transported by pipeline and tanker to markets. Similar GTL conversion processing could be accomplished on offshore platforms or barges in the Gulf of Guinea to facilitate gas and associated oil production from wells that do not have pipeline access.

Technologies/Projects For Consideration In The Niger Delta: Category A Projects: Natural Gas-To-Liquids:

Gas-to-liquids conversion technologies use chemical or physical means to convert natural gas to a liquid form suitable for ready transport or direct use. The conversion is accomplished in one of two ways:

- Compression and refrigeration, in which the gas is liquefied cryogenically and subsequently re-gasified for eventual use.

- Chemical conversion, in which the gas molecules are chemically altered and combined to form a stable liquid that can be used directly as a transportation fuel or petrochemical feedstock/product.

Chevron-Texaco is currently building a 30,000 b/d GTL plant in joint venture with Sasol of South Africa in Escravos that is expected to come on stream in 2005-2006. Chevron's GTL plant involves Fischer-Tropsch catalyst reactor and hydro processing / isocracking technology. It uses a modified slurry phase technique (used for coal in South Africa) and applied it for use with natural gas

There have been significant advances in GTL conversion technologies that significantly reduce costs and accelerate deployment for commercial application. Canada's Synergy Technologies Corporation have plans for a lower cost GTL plant for Nigeria and has opened an office in Lagos.

Synergy is working with the Nigeria National Petroleum Corporation (NNPC) and a local firm, Drake Oil on its plan for the 20,000 b/d plant. Synergy's low cost GTL technology is based on its SynGen reactor, which uses a cold plasma technique to create syngas. The second step in the process uses a hydrocarbon chain limiting catalyst in a Fischer-Tropsch reactor.

The chain limiting process, which is under license from its Russian developers, does not produce waxy products; so it produces clean, sulphur and aromatic free diesel and gasoline without the need for a hydro cracker. The capital costs are some 30% lower than other existing GTL processes.

Syntroleum and Brown & Root developed a GTL technology in partnership with Texaco (before the merger with Chevron) and is the first commercial GTL plant in the World to use hybrid multi-phase technology.

Other Syngas From Natural Gas Technologies:

The Transport Ceramic Membrane:

This process opens up the potential for air separation and partial oxidation of methane in a single-step ceramic reactor to make the intermediate product syngas, which can be readily converted to clean liquid fuels and other environmentally superior hydrocarbons. Syngas is composed of hydrogen and carbon monoxide, and is converted in a separate Fischer-Tropsch reactor to desirable long-chain, paraffinic hydrocarbons. The cost reduction using ceramic membrane technology for the chemical conversion is about 25%.

The prospects of this and other conversion process improvements enhance practical options for significant use of GTL fuel products (or blend stock) as a way to reduce air pollution and greenhouse gas motor vehicle emissions through better engine performance, without raising prevailing fuel costs. It has been shown that a significant in emission components can be obtained using syngas-derived diesel fuel (from gas and coal). Significant emission improvements have been noted in newer, better performing engines using cleaner GTL fuels.

Thermo-acoustic Natural Gas Liquifaction:

This is a new process for small-scale liquefied natural gas manufacture at remote offshore and onshore locations. This process uses direct gas burning to generate sound waves to drive a refrigerator. The process is designed for small-scale liquefied natural gas (LNG) generation at wellhead or other locations, at one-half of the cost of traditional refrigeration at similar scale. The process has no moving parts, does not require electricity, and lends itself for use at remote offshore locations. It could also be modified to produce liquefied petroleum gas (LPG).

This LNG process innovation has been taken from basic theoretical refrigeration concepts to a working unit capable of liquefying gas at a rate of up to 140 gallons per day at 60% efficiency. Following successful field-testing, it is possible to reach capacities of up to 1,000 barrels per day at efficiencies of 80% or better.

Small-Scale Methanol Plants:

Methanol can be produced through a gas synthesis process using small modular plants developed by HydroChem –USA. The HydroChem modular methanol plant offers a number of savings in plant cost. It is fabricated and assembled under in one space, with all components built of supplied in-house. HydroChem produces several gas synthesis plants each year and is the world's largest supplier of small gas synthesis plants with capacities in the region of 100-120 tons per day.

Shell Projects

When hydrocarbons are burnt in conventional ways emissions of carbon dioxide (CO_2) and sulphur dioxide (SO_2) are given off. Shell has developed the necessary technology to dramatically reduce these impacts, particularly for coal, centred on gasification. This involves partial combustion and is the key to a number of exciting new technologies that make it possible to use gas, coal and other hydrocarbons to produce ultra-clean fuels and electricity.

Shell gasification technology partially oxidises coal, gas (and even oil and biomass) to make syngas. Syngas is as clean burning as natural gas and when used in combined electricity generation, it produces 90 percent less SO_2 and 15 percent less CO_2 than conventional coal-fired plants. Syngas can also be processed to produce chemical feedstock, ultra-clean fuels or electricity using the Shell Middle Distillate Synthesis (SMDS) process. SMDS fuels are completely free from aromatic and sulphur components, leading to greatly reduced exhaust emissions.

Category B Projects: Coal-Based, Waste And Biomass Liquid Fuels & Chemicals.

The Niger Delta's coal reserves are of the low sulphur variety and are estimated at over 22 billion tons. While oil is currently Nigeria's major commercial energy source, the government intends to increase both natural gas and coal as a means of diversifying the country's primary energy mix. Ultra-clean transportation fuels, chemicals and carbon products can now available using technologies that can convert coal into liquid fuels and chemicals in two steps:

In the first step, coal is gasified in the presence of oxygen and steam to generate a gas containing mostly carbon monoxide and hydrogen (i.e. synthesis gas). In the second step, the synthesis gas, after being cleaned of impurities, is converted into a variety of products. These products include:

- Hydrocarbon fuels, such as gasoline, diesel fuel, and jet fuel.
- Oxygenated compounds, such as alcohol fuels (e.g. methanol), and oxygenated fuel additives (such as ethers and esters).
- Premium chemicals such as olefins and paraffin wax.

Advanced chemical processes are being developed that lead to greater process efficiencies and lower capital costs.

Projects involving coal gasification will focus on the production of clean fuels that:

1. Are environmentally superior to those derived from conventional oil-based fuels
2. Can supplement the liquid fuel requirements of the Nation's transportation fuels infrastructure
3. Will use the existing transportation fuels infrastructure
4. Will help engine and vehicle manufacturers achieve higher performance with significantly lower emissions.

The Liquid-phase Methanol from Coal Project:

This process for the production of methanol from gas is now a commercial reality and can be exploited for use of coal and natural gas in Nigeria. Also novel methods have become available to reduce greenhouse gases through process improvements and utilization of multiple feeds such as waste material or biomass.

Cost Implications:

With current technology, the cost of producing coal-derived fuels in stand-alone plants would be about $30 per barrel of crude oil equivalent

(COE). The cost can be reduced to $21 per barrel COE by co-production of liquid fuels and electric power. Novel three-phase slurry reactor technology has been developed to cost effectively produce premium fuels, an excellent diesel fuel-blending feedstock, or high-quality chemicals using syngas produced from natural gas, coke, refinery waste, and /or coal through the Fischer-Tropsch process.

Examples of Clean Coal Technologies:

Two Clean Coal Technology projects that addressed the conversion of low-sulphur coals into high-energy-density, very low-sulphur products in the U.S that can be adopted in Nigeria are:

ENCOAL's Liquids-From-Coal (LFC) process successfully completed test-run operations in July 1997,and has been commercialized. During the test-run, nearly 260,000 tons of raw coal were processed into 120,000 tons of solid process derived fuel (PDF) and more than 121,000 barrels of coal derived liquid (CDL), all using natural gas.

The US Western Syncoal Partnership's test-run of the Advanced Coal Conversion Process (ACCP) led to the production of high-energy-density, low sulphur solid Syncoal fuel. Through fiscal year 1999, the ACCP facility processed over 2.3 million tons of raw coal to produce over 1.5 million tons of Syncoal fuel.

Coal Quality:

Coal quality is an essential factor in coal gasification projects for the production of clean fuels and chemicals. It will therefore be necessary to establish a national coal-quality database on trace elements (in co-operation the Raw Materials Research Council).

Municipal Waste and Biomass:

Technologies have also been developed that combine coal and biomass or municipal waste into clean-burning fuels. Also developed is a process for removing mercury from coal before it is gasified, preventing the mercury from being released to form a hazardous air pollutant.

Projects Programme

What is proposed is the expansion of projects in the Niger Delta that uses the above technologies by establishing joint ventures involving the Nigerian private sector, Niger Delta Stake-holders, the oil companies and/or public sector agencies under PPP arrangements:

The Sustainable Technology Assessment Roadmap

This is an iterative analytical process that combines data, reports, stakeholder input, and industry intelligence in a common information platform. It uses a series of criteria selection screens to assess and sort relevant information from a variety of sources. The output is the highlighting of key technology investment opportunities for each sector

Sectors

Project focus can be on seven primary economic sectors:

Energy Exploration & Production – including Clean Conventional (oil and gas) and Renewable Fuels (bio-fuels, hydrogen production and purification). Note that Renewable Electricity and Renewable Fuels are linked as they share a number of technological platforms.

Power Generation – including Clean Conventional and Renewable Electricity Generation (wind, solar PV, bioelectricity and stationary fuel cells).

Energy Utilization – improving the effectiveness of the application of current end-use technologies in industrial, commercial and residential sectors (i.e. improving energy efficiency).

Transportation – including Systems Efficiency and Fuel Switching. Also note that Fuel Switching and Renewable Fuels are linked as they share a number of technological platforms.

Agriculture – addressing solid waste or Biomass conversion to Fuels and eliminating air and water contaminants produced by manure.

Forestry and Wood Products – addressing development of wood waste recycling technologies to harness energy resource potential, reduce emissions and improve productivity and profits.

Waste Management – addressing the various forms of waste management from municipal (and commercial) and primary and secondary industrial sources.

Investment Categories

There are three primary categories of investment opportunities:

- **Short Term Investment Priorities** – These are investments that could be made within the next 3-5 years that could have a direct and positive impact in the next 6-8 years.
- **Long-Term Investment Priorities** – These are early stage investments that could be made within the next 3-5 years but where the environmental impacts are realized over the longer term (greater than 8 years).
- **National Strategy Impacts** – Over the course of developing the SD Business Case a number of policy-related enablers and barriers to the development and implementation of sustainable technologies can be identified.

Assessment Descriptions

Once the market vision has been accepted, the economic sectors and their associated technologies are assessed through the following four screens:

Market

This focuses on the ability of the market to carry the emerging technologies that are currently at the development and demonstration stages. It identifies what needs to be done in order to maximize the application and acceptance of the technology, with a focus on financial and economic performance?

The main components of the assessment are:

- **General Market Description** – an overview of the sector under consideration, with a comparison to conventional or competing sectors.

- **Market Potential** – an indication of the immediate growth potential for the sector under consideration. The data is drawn from industry literature and stakeholder feedback, and shows the theoretical and realizable potential as well as equipment installed costs (where available). Using linear extrapolation, it then estimates the anticipated potential over the next three to five years. Due to the rapidly evolving nature of emerging markets, it is necessary to conduct this assessment a number of times as conditions change.

The primary purpose is to understand the gap between today's situation and the vision for each sub-sector.

This helps to determine the required rate of innovative developments and the amount and timing of capital placements.

There are three Market Assessment criteria used in the process:

- **Stage of Investment** – An assigned value (on a scale of 1-10) that takes into account market barriers, the amount of time expected for the technology to achieve full commercialization, market infrastructure issues and impediments, and current state of codes, standards and regulations.

- **Economic Efficiency** – An assigned value (on a scale of 1-10) that takes into account technology spin-off potential, product replicability and scale-up potential, market size and dynamics, competitiveness, pricing and financing, and export potential.

- **Emissions Reduction Potential** – A calculated value of the difference in GHG emissions between conventional technologies and the alternative technologies within the sub-sectors under consideration. It is shown in megatonnes of carbon dioxide equivalent (MtCO2e) and is the amount of CO2e expected to be reduced or displaced within the next three to five years as a consequence of commercializing the subject technologies.

45

Note that GHG is a proxy used as a general indicator of emissions reductions as, for most technologies, there is a positive correlation between GHG and other air emissions. Exceptions (such as the inverse relationship with NOx associated with combustion-based technologies) are noted where applicable.

The Market Assessment focuses on technologies at critical stages in the development cycle. Specifically, it is focused on those technologies that are between prototype development and market-ready product stages

Technology

This concentrates on the technologies that need to be brought to market in order to achieve the stated vision. There are 15 fundamental ranking criteria, which are weighted and rolled up into two principal impact criteria:

Economic Impact: The developmental and financial issues related to a specific technology that can/will influence sector growth, technological inter-dependencies, Infrastructure improvement, and the cost of environmental improvement; and

Environmental Impact: The magnitude of the emissions reduction potential, reductions of regional environmental pollutants, the life cycle emission returns, and the time at which these emissions reductions are most likely to occur.

Sustainability

This describes the impact that these technologies are likely to have on individuals, communities and regions. Each technology group is evaluated in terms of its potential impact in three key areas:

Economic – current investment capital, company and job creation, productivity impacts;

Environmental – impacts on wildlife, air (GGHG and regional pollution emissions), water and land; and

Societal – health and safety, training and education, and aesthetics and property value impacts.

Risk

This outlines the potential risks associated with the development and implementation of the technology, and are divided into three criteria:

Development Risk – will the technology work as intended?

Financial Risk – is there enough private capital to fully commercialize the technology and will it be financially viable once commercialized?

Market Risk – is there sufficient market demand and infrastructure to support the technology?

S I X

The Hydrocarbon Development Fund (HDF)

The proposed HDF is in recognition of the need for a venture capital fund for renewable energy projects in the Niger Delta and can provide the financial backbone for the National Clean Technologies Initiative (NCTI) It is thus proposed that this new outfit to be called the Hydrocarbon Development Fund Plc be owned by the Nigeria National Petroleum Corporation (NNPC), some Niger Delta State Governments, the International Finance Corporation IFC), the African Development Bank (ADB), Oil & Gas Companies, Private Nigerian Promoter/Investment Group, while at least 25% of its capitalisation will come from the Stock Exchange.

The Hydrocarbon Development Fund Plc (HDF) is a proposed new venture capital fund with the specific mission to support the growth of Renewable Energy projects in the Niger Delta region of Nigeria where because of new and exciting developments in the technology of sustainable energy production, providing innovative processes for electricity generation and environmentally superior clean fuels and chemicals

As previously stated and which is now obvious, the Niger Delta, as a strategic region of Nigeria, because of its Oil& Gas resources is in urgent need of substantial financial assistance to sustain economic activities in the area, most of which revolves around the energy sector. One way of doing this is to establish a special regional fund that can provide risk capital to fund Oil and gas field support services many of which fall into the medium scale in terms of size.

The connection between current energy use patterns and environmental degradation has been clearly documented, and such degradation will be severely accelerated in the future if new technological solutions and new priorities are not brought into the development planning process.

A different approach to energy that places greater emphasis on energy efficiency enhanced and increased commercial use of renewable forms of energy and the introduction of modern cleaner energy technologies in Nigeria in general and the Niger Delta in particular, is required.

The HDF-PLC will give priority to investing in the above projects including sustainable and renewable power generation for the manufacturing sector.

Tomorrow's generation of high-efficiency automobiles and trucks will need gasoline and diesel fuels that are environmentally superior to traditional fuels and can result in cleaner air and fuel efficiency (more miles-per-gallon).

This new generation of Liquid and Solid Fuels will include the following:

- Ultra Clean (low-sulfur) Transportation Fuels.
- Natural Gas – to – Liquids.
- Solid Fuels and Feedstock.
- Bio-fuels.
- Coal-Based Liquid Fuels and Chemicals.
- Sustainable and renewable power and steam generation.

The Hydrocarbon Development Fund PLC (HDF)

HDF can be a specialised financial institution comprising shareholders who operate within the oil rich Niger Delta and who have an interest in the supporting the creation, growth and development of Medium-sized Renewable Energy enterprises in that region which support the reduction of greenhouse gas emissions and a healthier environment for all its inhabitants.

Its Investment strategy will aim to equip the Niger Delta States with competitive economies, generating high quality job creation.

The HDF will create guarantee instruments to facilitate access to debt finance through the intermediation of a wide range of banks and financial institutions.

HDF will pursue community objectives and at the same time generate value for its shareholders. The HDF will always act on a commercial basis.

The multinational oil companies, selected banks and insurance companies, investment companies, the Niger Delta Development Commission (NNDC) etc, are being proposed as shareholders and partners in this new enterprise.

As much as possible HDF will appoint a non-executive director to the board of any beneficiary company, to ensure that management maximizes the value of the investment, give early warning signals of deviation from plans and ensure prompt attention to resolve identified problems. HDF will have a maximum investment portfolio of not than 20 companies at any given time.

After the first 12 months, it is recommended that HDF will apply to be listed on the Nigerian Stock Exchange to increase its capitalisation.

Proposed Equity Sources:

The NNPC Multinational Oil Companies operating in the Niger Delta, the NDDC, Bank of Industry BOI), Banks, IFC, Insurance Companies, Private Investors, the ADB, Investment Companies etc. These shares will be offered through private placement.

Mission Of the HDF

1. Globalisation will place an increasing demand on Nigerian companies to become more competitive, better-funded, more performance-oriented, product innovative and technology-

driven as the engines for growth of a modern emerging market economy.

2. The HDF will have a specific remit to support the creation, growth and development of and medium scale enterprises, particularly in the Niger Delta region. It will mainly intervene by means of equity and guarantee instruments.

3. HDF's shareholding will be offered to Multinational Oil Companies, Financial institutions such as the IFC, ADB, well-capitalised banks, Insurance companies, Investment Companies and the Niger Delta Development Commission (NDDC) etc. In order to raise additional capital it will seek an early flotation on the Nigerian Stock E

4. HDF will contribute to Nigeria's development objectives; in particular will be committed to the development of a knowledge-based society, centred on innovation, growth and employment, the promotion of entrepreneurial spirit, regional development and cohesion.

5. HDF will act independently and commercially under market conditions as it targets appropriate returns for its shareholders.

6. HDF's Guarantee Instruments will facilitate access to debt finance through the intermediation of a wide range of banks and financial institutions.

7. The guarantee instruments will be structured in a way that will reduce the risk exposure of the intermediaries providing debt finance to the medium sized companies. These guarantee products will be priced in accordance with the risk assumed.

8. HDF will be an Equity Fund based in Port Harcourt whose focus will be the provision of additional longer term financing coupled with "hands-on" management and marketing assistance.

9. HDF with permission of the of the Central Bank of Nigeria establish the an Equipment Leasing Facility to assist companies acquire production assets with greater ease.

10. HDF will play a catalytic role to attract private sector finance

11. HDF will optimise the impact and benefit of those operations in which it participates in close association with the financial sector and will contribute to the diffusion of best market practices.

12. HDF will be high volume supporter of the Nigerian venture capital market and act in close co-operation with the NNDC.

Strategic Objectives Of HDF:

1. HDF's main objective is utilising long-term equity capital for the financing of new ventures characterised by innovation, technological breakthrough and job creation. Medium-sized companies operating in the Niger Delta, focusing on the Oil& Gas, Renewable Energy and new oil & gas technologies, will be given priority attention.

2. HDF's investments will also focus as much as possible on "early stage" projects, but will occasionally provide funding for the rehabilitation of potentially high growth companies, management buy-outs and wealth creating ventures.

3. HDF will provide seed capital for financing the exploitation of research and development results.

4. Guarantee instruments will constitute the other financing tool at the disposal of HDF. The Guarantee Facility will cover equity guarantees

5. HDF will anticipate future activity in the Nigerian venture capital market by developing a capacity for analysis, be reactive and play fully its catalytic role.

6. Propose appropriate initiatives as far as new instruments are concerned, in close co-operation with its shareholders and partners.

7. Maintain a qualitative approach to its operations, based on a high level of staff expertise that seeks to diffuse best market practices

8. Transfer of know-how will be an important part of HDF's activity to the less developed regions of the Niger Delta. This

can be done though technical assistance, training and capacity building.

10. HDF will develop a high-level Communication Policy targeted at specialist circles.

HDF's ambition is to become a major financing vehicle for Renewable Energy projects in the Oil & GAS sector with Nigeria as the focus of its activities, hence the decision of its promoters to have a branch office in Port Harcourt in the heart of the Oil & Gas rich Niger Delta of Nigeria. HDF will also mobilise financial resources from the International Capital Markets to finance the development of oil and gas support companies in Nigeria.

HDF will also commit 30% of its resources to support oil field service companies providing support for:

* Greenfield/Expansion Projects
* Onshore/offshore (mature fields or proven reserves)
* Appraisal & development
* Oil & Gas pipelines
* LPG/LNG
* Refineries

Both domestic or foreign sponsors of projects will be encouraged; however:

- For New ventures or expansion, private sector majority ownership will be required
- Project must be commercially viable as well as developmentally sound.

Nigeria's Current Investment in Oil & Gas:

Current estimates of Nigeria's oil and gas reserves are about 22billion bbls of oil and 3.5 trillion cu metres respectively. However less than half of the countries basinal acreage has been subject to serious exploration. Current investment in exploration is concentrated in the offshore deepwater zone of the Niger Delta.

The Nigerian Government has a declared policy to increase current reserves to 40 billion bbls and daily production levels to 4.0million bbls by the year 2010. The oil and gas sector will remain the bedrock of the Nigerian economy for the foreseeable future.

The Nigerian National Petroleum (NNPC) has put in place a strategic investment plan, agreed with government. The Strategic plan covers both the joint venture operations, and NNPC own production.

Some elements of the plan include:

*   Increase of national crude oil reserves to 40 billion barrels
*   Stimulate and promote private sector investment in gas related industries
*   Ensure efficient and adequate supply of refined products throughout the country
*   Improve on rehabilitation and maintenance of all assets.

The Nigerian Government has introduced a number of incentives in an attempt to encourage new foreign and domestic private sector participation in the oil and gas sector

1. Tax Holidays:

A five-year tax exemption for new projects, from start of production

2. Equity Structure:

Foreign private investors can now own 100% of the shares in any venture incorporated in Nigeria

3. Guaranteed Export Earnings:

Earnings from the export of any local production will be permitted to be retained in approved export accounts in any country nominated by the investor.

SEVEN

Infrastructure

Energy, together with water, transport, roads and information & communications technologies (ICT) are the core components that form an integrated approach to the issue of rural infrastructure in the Niger Delta.

Rural Energy Priorities:

In the Niger Delta, very few of the rural dwellers, generally largely poor populations, dispersed in nature, have access to any form of modern energy such as electricity. In a country starved of access to efficient and reliable sources of energy, the benefits of energy supply are convincing if they are affordable.

Household energy in many parts of the Niger Delta is fuelled by biomass that is collected free of charge in the environment by household members (usually women and girls).

Commercially generated energy can only be a substitute at the households level if there is a willingness (and ability) to pay and if there are alternative income producing activities for household members who currently collect and burn biomass.

The following priorities need to be taken:

* The need to introduce appropriate, efficient and energy-saving production and processing technologies
* Accessibility, affordability and method in   which the energy is used are important.

The issues and challenges of rural energy up-liftment for the Niger Delta are many. However, through a systematic process of understanding the needs of the rural consumer, evaluating the life-cycle cost of each modern and innovative energy technology to package together a least-cost fit-for-purpose energy solution offered to the community, who have been consulted from the beginning, the platform will be set for sustainable development in areas currently untouched by modern technology.

In energising rural communities, urgent attention should also be paid to the following in fashioning an appropriate rural energy strategy for the Niger Delta:

* Human Capacity Development
* Productivity Improvement
* Rehabilitation of Infrastructure
* Reinforce National Grid
* Exploit Waste to Energy Opportunities
* Develop Greenfield Infrastructure
* Identify Synergies with other Utilities
* Extend Distribution Networks
* Exploit Renewable Energy Sources

Renewable Energy:

Solid biomass, such as fuel wood, is a major energy source for two-thirds of Nigerians especially in the rural areas. However this reliance on fuel wood to meet domestic consumption needs has resulted in deforestation in Nigeria. The Nigerian government intends to increase domestic coal consumption, partially in an attempt to slow deforestation.

In addition to increasing coal consumption, the Nigeria Energy Commission is initiating a community-based solar energy project aimed at providing photovoltaic power as an alternative to the National grid. This project aims to provide electricity to the 75%

of the population not serviced by the National Electric Power Authority.

The United States Department of Energy also supports the development of Nigeria's solar, wind and hydroelectric power sources. In 1999,the US government announced a project to deploy solar-powered water purification systems and other photovoltaic technologies to villages throughout Nigeria.

Wind, Ethanol and Biomass:

Give priority attention to the exploration and promotion of alternative energy sources for the rural communities especially wind energy through the establishment of wind farms, the exploitation of agricultural biomass energy and the production of ethanol as a by product of sugar cane molasses production in the abundant marsh swamps of the Niger Delta.

Water Infrastructure Services:

Government has for long recognised the urgent need to address the problems of water and sanitation in poor rural communities. Engagement of the private sector should be a key feature in reforms aimed at tackling these problems.

Multi-lateral development funding organisations and the international private sector should dialogue to seek a common approach to increasing funding flows to provide basic community services.

For this to be effective, a clear understanding of the potential of private contributions is needed. It is important to understand where the private sector is unable to assist as to identify areas and approaches where there are no alternatives.

Small-scale private providers, non-governmental and community-based organisations (NGOs and CSOs) have played a lead role in service provision to rural communities where public services have been inadequate. Their insights and experience in serving these communities and their potential contribution, as partners should be explicitly recognised.

Links between water supply and environmental sanitation in designing new programs in which the communities are consulted and carried along. Hygiene education becomes a crucial element to ensure a reduction in communicable diseases.

Transport and Roads:

Research into rural customer needs yielded two core challenges:

* The lack of an integrated approach to the provision of Infrastructure
* The absence of a framework for rural roads prioritisation.

The traditional design of infrastructure need to transcend its narrow scope to objectives that will ensure integration of farming communities through feeder roads, opening up of inland waterways through improved riverine and marine transportations and integration with oil and gas areas of activity to supply essential services.

Promotion of "spin-off" developments as propagated in the Spatial Development Initiative (SDI) approach should be applied to rural areas.

The Spatial Development Initiative (SDI) Concept

The implementation of spatial development initiatives can be regarded as a significant solution to integrating transport with broader socio-economic development particularly of the rural areas of the Niger Delta. The aim is to unlock inherent economic potential in specific Niger Delta locations.

The underpinning principles of SDI's are:

* Co-operation, collaboration and integration in terms of policy and strategy

* Focus on promoting development networks and not just roads alone
* Improvement of regional efficiency in terms of social services through integration and harmonisation.
* Increasing the role of the private sector
* Institutional collaboration

SDI's always link together a number of interdependent economic activities for the greater common good of all its components.

Some typical examples of SDI's are:

* "Agro-tourism" e.g. the Obudu Cattle Ranch & Tourism Resort in the Cross-River State in the Niger Delta.
* "Petroleum-from-Gas" e.g. The Chevron Escravos Gas-to-Liquids project and the new housing development for the Escravos Community whose citizens work or provide essential services for the project.
* "Eco-tourism" e.g. wild life holiday resorts
* "Techno-Manufacturing" e.g. a possible Silicon Valley or Internet Village in a specific rural area.

The strategic objectives of SDI'S are thus readily defined:

* To increase the tempo of rural economic growth and development
* To create a fertile seed bed for long term, sustainable employment in rural areas
* To promote economic integration and mutual economic support
* To grow export markets
* To correct in-balances in regional economic development and mobilise foreign investment.

E I G H T

Financing Road Infrastructure

Roads are an essential component of the transport system and one of the biggest investments in transport infrastructure. Any development strategy for the Niger Delta must address the road network as it relates to freight customers, rural customers, urban customers, tourism customers etc. There are two central issues emerging from the general state that we find our infrastructure in today. The two issues cover the sustainability problem and the deterioration of the road network in the region.

Road network deterioration

The most compelling overall finding was that of a road network in the region, in a serious state of disrepair to many that is non-existing. The priorities of the various Governments differ and are many, and this shift in priority can have the effect of major under-spending, affecting the networks quality, and creates additional costs in the system in the form of lower speeds, poorer safety, and increased vehicle wear.

Under-Investment in Roads

In addition, the situation reveals lower prioritisation of roads for investment of maintenance as a contributing factor to the sustainability problem. Most expenditure for roads goes to maintain the current system, rather than to implement a new strategic vision for roads in the region. There are critical funding shortfalls and lack of capacity at Government level. Still others find road investments competing unsuccessfully with other expenditures like education, health, housing and other social problems.

Financing Roads

The issue of road financing and pricing has been an important of economic analysis for a long time. For the past three centuries, it has been generally held that the funding of infrastructure should be tax-based i.e. through Government. Now we have entered the 21st century, infrastructure financing and pricing deserve a new and deeper analysis. We can no longer rely on Government alone to provide the finances. Given the pressing needs of Government to cater for the social needs of communities, while at the same time becoming more globally competitive, the climate has never been more conducive than now for the development of infrastructure using the pay as you use principle, rather than depending entirely on government.

Through innovative funding mechanisms and instruments, with the establishment of Public-Private Partnerships (PPPs), Governments in the Niger Delta could supply infrastructure within the constraints of budgetary limits. Projects funded by this method have shown potential to significantly stimulate job creation especially in the secondary labour market. A major benefit of this form of financing is that it offers Government the opportunity to obtain limited recourse funding (off-balance sheet) to provide infrastructure and services for sustained development and to meet political objectives. The method of funding is not a replacement for, but a supplement to government supported revenues from traditional sources and multilateral and bilateral loans.

Through appropriately structured concessions and franchises with adequate incentives to encourage project finance, government can leverage there available but limited direct financial and loan capacity resources. Further to the financial benefits, PPPs together with the corporatisation of various sectors of Government, will afford it the opportunity to reduce size of Government, increasing efficiency and effectiveness.

Additionally, innovations in financial technology and the globalisation of financial markets have introduced a larger pool of resources and a more diversified array of instruments that better match the financing needs of infrastructure projects.

The build, operate and transfer (BOT) concept can play a major role in the reconstruction and development programmes for the Niger Delta as it has done elsewhere in the advanced economies. The BOT mechanism and its variants enable Governments to obtain maximum leverage and ensure delivery without compromising fiscal integrity. Build, operate and transfer projects represent a viable solution to the Niger Delta's requirements to upgrade and expand infrastructure.

Transferring the responsibility for infrastructure provision from Government to the private sector implies more than just the transfer of ownership of the facilities, but also some or all of the decision-making authority, including the transfer of a sizeable amount of social responsibilities previously assumed by the Government. The widespread resort to PPPs in the Niger Delta means that Government reaps the fiscal advantages of reduced infrastructure provision. However, it becomes critical for Government to ensure that the funds released in this way are indeed strategically directed to meeting precisely those non-profit social needs that tend to be at best peripheral concerns for the private sector.

Optimal Economics for Road Construction

Considering the customer needs and optimal modal economics, suggests an asset configuration around the flows of goods and people in the region. These flows move along high volume corridors conveying freight and passengers. The goals are to recreate a system that has the lowest practical systems cost at the maximum affordability, with improved reliability and efficiency.

There are three key thrusts, namely:

- High volume corridors.
- Sustainable operations.
- Improved efficiency.

The logic of high volume concentration to drive down cost and improve service underpins the vision. Higher volume increase capacity utilisation of the corridors and lowers unit cost, while the integration of the corridors and their feeders further reduces cost and increases responsiveness to customer needs. Flows within corridors must be enhanced and sited at the point of best economic utilisation.

Provided that the funds available to Government are directed to the core needs, it is likely to be self-sustaining, while at the same time, working to upgrade and extend services. Any other road projects outside the core requirements will have to be funded by user revenues and self-generated cash flow. Government should give clear priority to those who have no access to the system for reason of affordability or geography. The incentive to expand into areas within the defined corridors must be there.

The benefits of this strategy are as follows:

- Greater value for customers.
- Improved profitability for business users.
- Lowering of the fiscal burden on Government.

Strategic Actions

The strategy programme for roads in the Niger Delta, must cover four principal streams of action:

- *Align roads strategy with customer strategies (industrial, agricultural, tourism, rural development, urban etc)*. The roads network is comprised of a combination of roads, owned and operated by the Federal Government, State Governments, and local Governments. Users from every different customer group travel on the roads. The road investment and maintenance strategy should be oriented to the strategies of each of the customer groups. Thus for freight, road investment must be aligned to the corridor strategy. Similarly roads for tourism should follow the tourism strategy, roads for agriculture to follow the agricultural strategy and rural roads investment should follow the rural development and prioritisation framework for funding roads in sustainable communities.

63

- Focus Roads Investment: Although all customers desire comfortable speed in dual-lane carriageways for their long distance travel, there can and must be the focusing of scarce resources on the core networks that reflect economic segments/clusters. The prioritisation should also distinguish economic roads (e.g freight, tourism and urban corridors) from social roads (e.g development roads, dedicated rural roads etc). Increase road funding to the priority road networks and differentiate spending by level of priority.

  Increase the overall level of funding to the network to invest in better roads. Early re-investment prevents higher operating costs and higher maintenance costs later. Do not actively invest in upgrading or expanding urban roads that serve mainly cars.

- Create Institutions to Support a Sustainable Roads Network: There is a clear need for a single institution/agency to coordinate road investment across all jurisdictions, so that Federal Government, State and Local Governments can uniformly prioritise and fund road construction and maintenance along key passenger and freight corridors. This will ensure that customer needs are met and that funds spent on planning, maintenance and construction are used as effectively and efficiently as possible.

N I N E

River Transportation

River transportation services for the coastal communities of the Niger Delta region should be a major plank of any strategy of achieving accelerated sustainable development of the Niger Delta region of Nigeria. The proposed Infrastructure project will have significant medium term impact in reducing prices of essential agricultural products and other food items in the region as river transport is a key form of transport for the distribution of these items in the region.

There should be a coordinated approach for improving passenger marine transportation services, particularly for the river communities in the Niger Delta by the local construction and supply of modern passenger boats and establishing an effective infrastructure for the efficient running of the service and maintenance and repair of the vessels.

The Cabotage Act 2003

On 30th of April, 2003,the Federal Government of Nigeria (FGN) signed into law an Act to restrict the use of foreign vessels in domestic coastal trade, cited as the Coastal and Inland Shipping (Cabotage) Act 2003. This act came into force Ist May 2004

Act Requirements and Implications:

The Act covers the Nigerian Economic Zone and Inland Waterways. The Act requires that vessels be built in Nigeria, wholly owned and manned by Nigerian Citizens.

The National Maritime Authority Register of Ships (ROS) shall duly register every vessel intended for use under this Act

Vessels eligible for registration under this Act are passenger vessels, crew boats, bunkering vessels, fishing trawlers, barges, off-shore service vessels, tugboats, anchor handling tugs, supply vessels, floating petroleum storage, dredgers, tankers, carriers and any other craft or vessel used for carriage on or through or underwater of persons, property or any substance whatsoever.

Based on the above we are proposing the establishment of a shipyard in the Niger Delta to build passenger vessels (specifications below) initially, then other types of vessels later. The expertise of experienced foreign marine companies can be utilized in executing the project. To support vessel flexibility, a new Shipyard can be established in the region to provide new vessel construction, conversion and repair services on the Intra-coastal Waterways.

Such a facility in the Niger Delta region of Nigeria that will be jointly owned by NDDC, State Governments in the region and Private Interests and structured as a Private Public Partnership (PPP) The goal will be to provide new construction services including the designing and building of passenger vessels initially and the eventual building of all types of vessels.

Repair and refurbishment services can feature 4 floating dry docks, capable of building and repairing marine vessels up to 15,000 HP. Another component of the operation can be a full service machine shop with specialty equipment best suited to the marine transport and oil industry in the Niger Delta.

Community-based Transport Services Managed by the Communities:

It is proposed that each community selected for the service will set up it's own structure to run and manage the service on a commercial basis. Based on information available to us we suggest implementing this proposal in 2 phases:

Phase 1 (First 18 months)
Rivers, Delta and Bayelsa States:

In each of the above States, 3 communities/transport routes will be identified. Each community/route can be supplied with 8 boats as follows:

*   4 vessels of 50-passenger capacity.
*   2 vessels of 25-passenger capacity.
*   2 vessels of 15-passenger capacity.

     Total number of vessels -per State: 24

Phase 2 (Second 12 months) - Akwa Ibom Cross-River Edo States Coastal Corridor Ferry Services

Given the coastal geography of Niger Delta coastline Africa, transport by sea is an ideal and relatively safe alternative to other modes of transport. Particularly a six-day trip along a coastal corridor connecting Warri, Port Harcourt, and Calabar offers a price effective, convenient and comfortable method of travel.

Warri Port Harcourt and Calabar represent three key economic nerve centres on the coast of the Niger Delta and thus patronage of this unique ferry service should pose no problem considering the growing economic importance of these port cities in the Niger Delta Transport by sea in an appropriate vessel enables a range of fares to be charged, from a premium luxury class, down to a low budget class.

Transport by air is more rapid and convenient, but services are becoming overstretched and as a consequence variable in performance. A high profile ferry service between the above ports will serve both the Oil & Gas and tourism sectors of the regional economy, attracting potentially substantial demand from the business and tourism communities.

For a new commercial venture it is important to minimize the financial risk to the shareholders. Thus, although the prospects for ferry services in Africa are very good, it is prudent to commence operations with a low breakeven point.

Once the ferryboat service is well established, the substantial cash flow generated from its operations can be used towards the purchase of new ferryboats

T E N

Energy Development in the Niger Delta

Globalisation of the World economy, including the elimination of barriers to trade and investment and the recognition that market forces promote the most efficient allocation of global resources offers great promise for Nigeria and Africa generally, to accelerate development and promote the well being of its citizens.

Modernization and industrialization of Nigeria however, will also imply increasing demands upon the energy sector. Energy generated in the Niger Delta can not only meet its regional demands but could become the most important energy source for the whole country and the sub-region.

Current Energy Issues:

The Niger Delta needs a conceptual integrated energy plan identifying key linkages with sustainable human and economic development, these being:

1. Access to affordable Energy- Empowering Niger Delta people
2. Reliability and Security of Supply- Sustaining accelerated economic growth
3. Environment-Preserving Niger Delta's Heritage
4. Institutional Arrangements and Funding Equitably Sharing through Partnerships
5. Planning-Defining Niger Delta's future Energy requirements
6. Implementation-Realising 'Goals

Energy Vision:

The proposed Energy Vision for the Niger Delta should be:

The Niger Delta with a business and policy environment that enables the effective delivery of energy services to its peoples and the economy at large, to support accelerated and sustainable development and economic growth.

Energy Scenarios and Resource Estimates:

It should be aspired that by the year 2020:

a) About 70 percent of Niger Deltans should have access to a reliable supply of energy from either central or distributed generation facilities

b) The Energy supply industry will enable an economic growth of 6 percent over the next five decades

Energy Priorities:

In order to realise the energy scenarios in terms of developing Niger Delta's energy infrastructure the following priorities are essential:

* Institutional Reform
* Human Capacity Development
* Productivity Improvement
* Rehabilitation of Infrastructure
* Reinforce National Grid
* Develop zonal/regional pools
* Energy/Power Market Modelling
* Establish Co-ordination Fora
* Reinforce Transmission Lines
* Exploit Waste to Energy Opportunities

* Develop Greenfield Infrastructure
* Identify Synergies with other Utilities
* Extend Distribution Networks
* Apply Demand Side Management Programs
* Energise Rural Communities
* Exploit Renewable Energy Sources

Funding available from multilateral and bilateral agencies will not be sufficient, thus highlighting the need for private participation.

To encourage private investment, Nigeria should be seen to be addressing the institutional arrangements that currently hamper private sector investment, while proactively promoting and participating in identified feasible projects, thereby ensuring early successes that would counter the current perceptions of risk, which can lead to escalating investment.

Gas Infrastructure Development:

The Niger Delta is richly endowed with natural gas resources estimated at over 159 trillion standard cubic feet, the equivalent of 25 billion barrels of crude oil. Nigeria has the 9th largest gas reserves in the world and approximately 30% of African gas reserves.

These gas reserves are expected to increase significantly as a result of the recent discoveries in onshore and offshore exploration and development. These reserves can handle present and projected demand for gas utilisation.

Much of this gas is associated gas, as many Nigerian oil fields are saturated and have primary gas caps.

It is estimated that about 51% of the associated gas (2 billion standard cubic fee of gas) is currently being flared in Nigeria, the highest among OPEC countries. About 12% is re-injected to enhance oil recovery.

Efforts to reduce the volume of gas flared are starting to reap dividends, with projects such as the Rivers State LNG facility; the Escravos gas gathering plant, the Chevron/Sasol Gas-to-Liquids Plant in Escravos, the NGL Plant in Bonny, the establishment of Shell Gas Nigeria Ltd etc.

The main thrust of Nigeria's Gas Policy is the elimination of gas flaring by 2005.The recent Production Sharing Contracts signed with the various oil companies awarded new blocks, now include gas utilisation clauses.

Gas producers are to carry out gas field optimisation studies on their respective concessions while the government agency NAPIMS would be responsible for overall optimisation planning of gas field development.

Incentives are also offered under the Associated Gas Utilisation Fiscal Incentives Decree, in an effort to put in place investment required to transport gas to interested third parties.

Critical to the development of a domestic gas market, is the establishment of a national gas grid system. This will require the construction of pipelines and depots to support a national supply system.

The National Gas Policy places the responsibility for the transmission system on the private sector. Utility companies are required, who will take gas from established city gates, and will then service private users.

The gas will be bought from the Nigerian Gas Company (NGS) on signing a sale and purchase agreement. NGC is the NNPC subsidiary with overall responsibility for gas development in Nigeria.

The Need For A New Rural Energy Policy:

I am proposing the preparation of a new Rural Energy Policy for the Niger Delta that can be implemented by the various State Governments in the region conduct.

Energy, together with water, transport, roads and information & communications technologies are the core components that form an integrated approach to the issue of rural infrastructure.

Rural Energy Priorities:

In the Niger Delta very few of rural dwellers, generally largely poor populations, dispersed in nature, have access to any form of modern energy such as electricity.

In a country starved of access to efficient and reliable sources of energy, the benefits of energy supply are convincing if they are affordable.

Household energy in many parts of the Niger Delta is fuelled by biomass that is collected free of charge in the environment by household members (usually women and girls). Commercially generated energy can only be a substitute at the households level if there is a willingness (and ability) to pay and if there are alternative income producing activities for household members who currently conflict and burn biomass.

The Policy document on completion will go a long in attracting private investment to compliment Government efforts and improve the overall energy infrastructure in the state.

Rural Energisation of the Niger Delta:

Millions of households in Niger Delta remain off the National Grid, many of them in remote areas where off-grid technologies offer the only practical means of supplying power.

The cost of reaching these areas by extending the grid is often prohibitive; the average cost per connection in urban areas is about US$285,while in the rural areas the figure rises to between US$500 and US$571.This means that alternative ways must be applied to service the rural areas, particularly in the field of non-grid technologies.

Significant opportunities are now emerging in Nigeria as the conclusion of restructuring/privatisation of the electricity sector and as government considers new technologies for off-grid service. State Governments in the region must develop a strategy for off-grid electrification and put in place non-grid electrification schemes to service remote rural areas.

Integrated Rural Energy Development Strategy For The Niger Delta:

This should form part of an Integrated Sustainable Rural Development Strategy (ISRD) designed by government with professional experts, to facilitate and co-ordinate service delivery to rural communities. Government should seek the involvement of the private sector, which is the basis of our proposal. Selected private sector companies can be given 20-year concessions by the Federal Government working with State Governments to implement the schemes.

Rationale:

Nigeria has approximately 5,900 megawatts (MW) of installed electric generating capacity, in the form of three hydro-based stations and five thermal stations. Nigeria is in the process of expanding electric generation, transmission, and distribution systems, with the long-term goal of reaching 25,000 MW in generating capacity.

Currently, only 10% of rural households and approximately 40% of Nigeria's total population have access to electricity. NEPA plans to boost this share to 85% by 2010.NEPA'S plan would call for an additional 15,000 kilometres (9000 miles) of transmission lines, 16 new power plants, and new distribution and marketing facilities.

The Nigerian Energy Commission and the Solar Energy Society of Nigeria are working on the implementation of a solar power system to meet the needs of rural villages and communities not served by the National power grid.

73

The Nigerian government is encouraging the increase in foreign participation in electric power through independent power plants that will generate and sell power.

Historical evidence clearly indicates that access to the services that modern energy provides is a pre-requisite for sustainable social development and economic growth.

In addition to increasing coal consumption, the Nigeria Energy Commission is initiating a community-based solar energy project aimed at providing photovoltaic power as an alternative to the National grid. This project aims to provide electricity to the 75% of the population not serviced by the National Electric Power Authority.

The United States Department of Energy also supports the development of Nigeria's solar, wind and hydroelectric power sources. In 1999,the US government announced a project to deploy solar-powered water purification systems and other photovoltaic technologies to villages throughout the Niger Delta.

Non-Grid Energy Generation for Remote Areas:

This proposal centres on the provision of non-grid electricity to households in remote communities in the Niger Delta region of Nigeria using the abundant flared natural gas to drive micro-turbines and fuels cells.

There have been dramatic technological breakthroughs in the fields of fuel cells and micro-turbines. Fuel cells are devices, which work by combining hydrogen with oxygen from the air to produce electricity. The pace of development of fuel cells was dependent on the oil industry agreeing on what fuel to use as a source of hydrogen. Ideally pure hydrogen gas is the best source. But oil companies argue that this requires a lot of storage capacity and would mean building a new fuel infrastructure at vast expense.

It has now been shown that fuel cells can utilise natural gas; reacting it with steam to release the hydrogen in it first reforms the natural gas. Reformation eliminates the need to supply to supply the fuel cell with expensive hydrogen, making the whole process cheaper.

The Niger Delta Development Commission (NDDC) can, with the assistance of NGO/consultants develop a strategy for off-grid electrification and put in place non-grid electrification schemes to service rural communities in the Niger Delta.

The NDDC can seek the involvement of the development agencies (such as the UNDP, USAID and the World Bank) and the private sector, which is the basis of our proposal. Selected private sector companies can be given 5-year concessions working with NGOs/Consultants and Local Governments to implement the scheme.

The strategy for off-grid electrification could commence with the identification of potential areas for implementation of a model. Four (4) or five (5) specific areas (concessions) could be chosen and given out to 4-5 private sector companies (concessionaires), as a trial, to supply off-grid services in the specified rural areas.

The concessionaires will operate in the specified areas under the monitoring supervision and co-ordination company that can design the complete program for government.

After signing the concession contracts, the concessionaires will be required to develop a detailed business plan that will set out technical and commercial strategies, proposing to government the subsidy required for the financial viability of the programme.

Once government receives these plans, the levels and conditions of subsidy can be determined.

The scheme and strategies will rationalize rural electrification and contribute to the delivery of services to remote areas of Nigeria.

The Use of Methanol:

The other source of hydrogen for the fuel cell is by reforming methanol (which can also be produced from natural gas).

Industry observers are more confident about the market for stationary fuel cells. Allied Business Intelligence, an independent US researcher believes that global electricity generating capacity from fuel cells will grow to about 15,000 mw by 2010.

Micro-Turbines:

Micro-turbines are ideally suited to natural gas. Micro-turbines are seen as an intermediary step towards fuel cells, allowing efficient small-scale generation of electricity from a variety of more traditional fuels, including natural gas, propane (LPG), diesel, kerosene and methane based gases from waste. The micro-turbine has only one moving part. This is a high-speed compressor-cum-motor that spins at up to 100,000 revolutions a minute. The near absence of moving parts means that micro-turbines are cheap to operate and maintain, costing as little as a third of the running costs of a comparable diesel generator.

A version of a micro-turbine has been developed that uses sophisticated air bearings that require no liquid lubrication. Sizes range from 25kw up to 500kw. In developing countries like Nigeria, it costs between $1,000 and $1,500 per kW to build or replace electricity grids. Micro-turbines are thus an attractive alternative. International agencies such as the World Bank, as well as private-sector operators and non-governmental groups are devising micro-finance schemes to help bring electricity to low-income groups in countries like Mongolia and India and is thus also possible in the Niger Delta. Cost of Micro-power has come down to economic levels and trends suggest they will fall still further over the remaining decade.

E L E V E N

New Gas-Based Industries

A key policy thrust of the Nigerian government is to encourage and promote the establishment of infrastructure by the private sector, for the production of clean fuel products for export and local consumption.

Proposed Niger Delta Methanol Production Programme (NIMEP)

Interested promoters can seek the support of the US Department of Trade (USTDA) and the World Bank Group to assist in the funding of the necessary Feasibility/Technical studies for these projects that can start with the proposed the Onne EPZ Methanol Project in Rivers State, Nigeria.

This project is in line with and supports the objectives of the Global Gas Flaring Reduction (GGRF) Programme being pursued by the Federal Government of Nigeria in association with the World Bank. The Nigerian Government has set the target of ending all flaring by 2008, and has endorsed the Voluntary Standard for Global Flaring and Venting Reduction developed by GGFR.

The GGFR, whose global office is in Washington, DC20433, USA is a forum of governments of oil-producing countries, state owned oil companies, international oil companies, as well as other key stakeholders, led by the World Bank Group, which supports the efforts of the petroleum sector worldwide to progressively reduce emissions related to crude oil production.

Part of the GGRC objective is to identify potential areas/project categories for associated gas use (with special attention to gas-to-power and gas-to-liquids for the export market.).

A Proposed Five-Year Rolling Plan:

The aim of the 5-year rolling plan is to facilitate and participate in the establishment of four (4) methanol plants in the Niger Delta region of the country that will use flared associated gas as feed-stock and thus a make a significant contribution to the reduction of gas flaring, while earning substantial foreign exchange for the country and shareholders alike. All the 4 projects will be structured as PPPs and are as follows:

- Ist 2 years: The Onne EPZ methanol project. Although the plant will be land-based, the associated gas will come from both offshore and onshore fields. This project is described in more detail in this report.

- Ist 2 years: The Bayelsa methanol project. (to be established simultaneously with the Onne project). The actual plant will be installed on an offshore platform, close to the source of associated gas , offshore Forcados. The methanol produced will be transported by pipeline to the Forcados Export Terminal facilities for loading into special methanol tankers that will take the product to US and European markets.

- Next 3-5 years: Will see the establishment of two offshore platform based methanol projects in Koko in Delta State and in Qua-Iboe in Akwa Ibom States. Like the Bayelsa project both locations were chosen because of the availability of nearby export terminals for loading the finished product onto special tankers for export to the US and Europe.

- The promoters will look into the possibility of outsourcing methanol production to Nigeria by transferring the idle plant of US methanol producers (that have closed operations) to Nigeria, that has the advantages of cheap natural gas feedstock, relatively low labour and overhead costs and special incentives for gas-based industries.

The proposed production plant will be located at the Onne EPZ, in near Port Harcourt, Rivers State of Nigeria and will convert flared and captive natural gas from fields in the State into commercial grade methanol for export to the U.S.A and world markets. When operating at full capacity, the plant will monetize over 200 million cubic feet per day of natural gas from the off-shore and on-shore fields in Rivers State, to produce 2,500 metric tons (19,880 barrels per day) or more of methanol per day for commercial markets.

High US gas prices have boosted methanol prices and shut down most US producers. In addition, the promoters can charter two 45,000-dwt methanol tankers from Mitsui OSK Lines. These ships will carry the methanol to customers mainly in the US, but also in Europe.

Gas Development In Nigeria:

Nigeria is richly endowed with natural gas resources estimated at over 159 trillion standard cubic feet, the equivalent of 25 billion barrels of crude oil. Nigeria has the 9th largest gas reserves in the world and approximately 30% of African gas reserves. These gas reserves are expected to increase significantly as a result of the recent discoveries in onshore and offshore exploration and development. These reserves can handle present and projected demand for gas utilisation.

Much of this gas is associated gas, as many Nigerian oil fields are saturated and have primary gas caps. It is estimated that about 51% of the associated gas (2 billion standard cubic fee of gas) is currently being flared in Nigeria, the highest among OPEC countries. About 12% is re-injected to enhance oil recovery.

Efforts to reduce the volume of gas flared are starting to reap dividends, with projects such as the Rivers State LNG facility; the Escravos gas gathering plant, the Chevron/Sasol Gas-to-Liquids Plant in Escravos, the NGL Plant in Bonny, the establishment of Shell Gas Nigeria Ltd etc.

There is also the Mobil/NNPC joint venture of the Oso NGL (Natural Gas Liquids) project, which produces 110,000 bbls/day of NGL from condensate discovered in 1992 in OSO in the Akwa Ibom State of Nigeria. Reserve estimate of NGL, is around 350Million barrels.

The main thrust of Nigeria's Gas Policy is the elimination of gas flaring by 2008.The recent Production Sharing Contracts signed with the various oil companies awarded new blocks, now include gas utilisation clauses. Gas producers are to carry out gas field optimisation studies on their respective concessions while the government agency NAPIMS would be responsible for overall optimisation planning of gas field development.

Incentives are also offered under the Associated Gas Utilisation Fiscal Incentives Decree, in an effort to put in place investment required to transport gas to interested third parties.

Critical to the development of a domestic gas market, is the establishment of a national gas grid system. This will require the construction of pipelines and depots to support a national supply system. The National Gas Policy places the responsibility for the transmission system on the private sector. Utility companies are required, who will take gas from established city gates, and will then service private users.

Feedstock:

Associated natural gas from the Nigerian Oil fields in Rivers State will serve as Feedstock for the methanol plant under a purchasing agreement with the SPDC (Shell). It is envisaged that initially, the plant will process approximately 220mcf(million cubic feet) per day of associated natural gas as its main feedstock.

The Onne Oil & Gas Export Free Zone:

The exclusive zone for oil & gas business activities came into existence with the enactment of Decree 8 of March 1996 to actualise government's objectives of:

* Attracting foreign Investment
* Bringing about capital growth

* Encouraging technology transfer and skills acquisition
* Creating employment opportunities
* Boosting revenue generation for those projects established in the zone.

It is one of the fastest growing world gas zones dedicated to Oil & Gas and managed by the Onne Oil & Gas Free Zone Authority and DMS Nigeria Ltd have a developed infrastructure already in place and also contains a number of other support facilities including:

* Fabrication yard
* Dry Dock
* Transit and marine shore-based services including support vessels, barges and tug-boats* Ancillary/Logistic services

By locating this gas project within the zone, reduces the required initial expenditure by about 15%. Some of the incentives include:

* Exemption from import duty and excise duty.
* No value-added tax and withholding tax chargeable on free zone storage facilities and services.
* No corporate or personnel income tax
* No expatriate quota and residence permit required for expatriates working in the zone.
* No pre-shipment inspection and issuance of bank form M required prior to shipment of materials from country of origin etc.

New Gas-to-liquids Project at the Proposed Warri EPZ

This will be the first private project of its type that will utilise modern gas-to-liquids to produce synthetic fuels in Nigeria that are sulphur and particulate free. *A group of domestic and foreign airlines operating in Nigeria will be invited to participate in the ownership of the project since one of the products of the project will be JET-A FUEL. This will lead to a significant reduction in their operating costs emanating from the high cost of Jet Fuel*.

Feedstock:

Associated natural gas from the Nigerian Oil fields in Delta State will serve as Feedstock for the project under a purchasing agreement with the SPDC (Shell). It is envisaged that initially, the Warri GTL project will process approximately 220Mmcf/day of associated natural gas as its main feedstock.

Natural gas liquids (NGLs) will be stripped from the stream for further cracking to produce high quality liquid fuels including Jet Fuel, for domestic consumption and export, while the remaining LPG will be for the domestic market. Liquefied Petroleum Gas (LPG) is hydrocarbon from oil bearing strata which is gaseous at normal temperatures but which are liquefied by refrigeration or pressure (compression) to facilitate storage or transport. They are mainly propane and butane and used domestically for as cooking gas.

Product Range:

The Warri GTL project will produce high quality liquid fuels and associated products from natural gas and condensate found in abundance in the Niger-Delta region and propose to enter into some joint arrangements with any of the major oil producing companies currently operating in the region for utilising their gas.

The technology can be procured from South Africa through a tri-partite arrangement with a South African company utilising the technology but wishing to diversify their operating base by entering into a strategic alliance with Nigerian parties to utilise the abundant Nigerian gas reserves for a new Nigerian gas processing project.

The South African company will be given the exclusive technical management responsibility to run the plant efficiently and profitably. Talks may also be held with the Nigerian Gas Company. Specifically the proposed Warri GTL project will produce petrol, diesel, Jet fuel, liquid petroleum gas (LPG) and other products. Over the long term, the business vision of project owners will be to add value to its operations as studies have shown that there is substantial potential for the production of petrochemical derivatives such as propylene, ethylene and butylene.

The execution and implementation can be carried out in stages starting with the incorporation of an offshore company and two local Nigerian companies.

Investment Incentives:

This project follows the present government policy to encourage foreign and domestic investment by giving the following incentives:

i)   Tax holiday of 5-12 years from the start of production.

ii) Gas or Crude supply to be guaranteed at prevailing international prices.

iii) Guaranteed export earnings and banking incentives-overseas escrow accounts can be used for investment transactions and repayment of the construction loan.

iv) Customs Duty Reduction (50%)- on equipment, plant and materials for the construction of the refinery.

Project Description-Summary

i)   Liquefied Natural Gas and condensate into high quality, environmentally-friendly liquid fuels namely, petrol (PMS), diesel (AGO), JET-A Fuel, Liquid petroleum gas (LPG) and fuel Oil (LPFO, HPFO) as well as a range of anhydrous alcohols.

ii) The other train converts crude oil into similar products on a lower scale and in addition produces petrochemical naphtha and lubrication oil.

All products will meet or even surpass NNPC/US (API) and international standards.

By locating this gas project within the zone, reduces the required initial expenditure by about 15%.

Some of the incentives will include:

* Exemption from import duty and excise duty.

* No value-added tax and withholding tax chargeable on free zone storage facilities and services

* No corporate or personnel income tax

* No expatriate quota and residence permit required for expatriates working in the zone

* No pre-shipment inspection and issuance of bank form M required prior to shipment of materials from country of origin etc.

Markets:

55% of products will be produced for markets in South Africa, USA and Europe. 45% of its production however will be for domestic consumption, while the petrol will be for the South African and West African markets. The diesel, kerosene/Jet fuel, naphtha and fuel oil will be for the US and European markets.

TWELVE

Industrialising The Niger Delta

The previous government's liberalization and economic reforms programme was aimed at substantial economic growth and a harmonized integration with the global economy. The changing global and domestic environment requires reorientation of the development vision, especially in the Niger Delta region in recognition of the new paradigms that should now focus on industrial policy initiatives as an important integral part of the overarching economic development of the nation.

Development efforts in the region so far have focus on infrastructure led growth. Any new industrial policy initiative should seek to consolidate the past progress and lay the foundation a vibrant Niger Delta economy that focuses on improving the quality of life of its residents through industrialisation. .

Industrial policy should aim at promoting rapid industrial growth in the region by creating an investor-friendly environment that facilitates industry moving strongly to the front ranks of global competition. The policy should further seek to promote integration of private sector initiatives in the industrial development process of the Niger Delta.

It should aim at promoting industrial growth in the context of overall economic development in the region by creating an investor friendly enabling environment that facilitates the industry to move strongly to the front ranks of global competition.

A new industrial policy should have the following objectives:
- To fast track and modernize industrial development in the Niger Delta in line with global trends.

- Attract new investment and growth of existing industry.

- Increase employment in industrial and allied sectors of the economy by 20% over the next five years.

- To attain sustainable economic development through catalysis of investments in all sectors of the regions economy.

- To achieve larger value addition within the region thereby contributing to a higher quality of life for its residents.

These objectives can be achieved by:

- Adopting a co-ordinated approach to the development of all sectors of the economy in the region that comprehensively addresses economic value addition.

- Focusing on creation of enabling mechanisms for higher levels of co-ordination between the various arms of government.

- Revamping the institutional processes and reorienting their role to ensure fast track clearances and an investor friendly environment.

- Simplification of rules and procedures by adopting a facilitation role that enables smooth and successful operation of industries.

- Instituting an effective Monitoring and Grievances Redressal mechanism.

- Modernisation and simplification of administrative procedures to start and operate business in the state.

- Execute a planned development of human resources through greater co-ordination between industries and technical institutions generating higher employment for its residents

- Focusing on small and medium enterprise (SME) renewal and channeling new investment into emerging sectors of the regions economy.

- Promote the growth in the services sector, recognizing its critical role as the driver of future economic growth.

- Excercising fiscal prudence in prioritizing public investment and through reforms in taxation.

- Rationalise and revise the package of incentives making it more effective and meaningful for the speedy development of the region.

Economic Development Board (EDB)

Each State Government in the region should establish an Economic Development Board (EDB) under the Chairmanship of the Deputy Governor and will include representation from the private sector in the individual States. The EDB will provide policy directives and oversee implementation of the sectoral components of an economic development plan of their states for achieving economic value addition through coordinated of Agriculture, Industries, Services (eg Tourism) Infrastructure etc.

Revamping Institutional Mechanisms

An Empowered Committee should be constituted under the chairmanship of the Secretary to the Federal Government to finalise and drive the new industrial policy initiative for the region, to monitor implementation of the new policy and co-ordinate with the various state governments in the region to achieve the objectives set forth in the policy.

An Industrial Assistance Group (IAG) should be set up and suitably strengthened and equipped to act as the nodal institution for providing assistance to investors in the region, escort services for venture location, information on investment policies (including incentives), procedures and clearances.

The Federal Government should constitute a Standing Committee representing all stake-holders in the region, to study the existing laws and procedures relating to setting up of new industries and suggest modifications and alterations that eliminate delays and expedite clearances. This Committee can also recommend amendments to and deletion of various laws and enactments that have outlived their usefulness.

Modernisation of Administrative Procedures

The Federal Government will encourage each of the State Governments in the region to introduce modern management practices in their ministries through extensive use of information technology to concurrently change its administrative procedures in line with modernization of the private sector. Transaction automation and information databanks should be created to bring about efficiency, accountability and transparency in administration. The IT plan of each State Government in the region shall be prepared covering critical areas and implemented in a time bound schedule.

Facilitation through Simplification of Rules & Procedures.

State Governments in the Niger Delta should adopt a system of deemed clearance/approval for speedy implementation of projects. Where provision for deemed clearance exists, Single Window Service (SWS) should issue the certificate of deemed clearance for projects having a Fixed Capital Investment (FCI) up to N30 Million. For projects having FCI of over N30 Million, such projects should have their certificates issued by the Industrial Assistance Group (IAG). There should be a time schedule fixed for giving necessary approvals/sanctions to reduce the time frame for project completion.

Grievances Redressal Mechanism

The Single Window Services (SWS) will be headed by an experienced civil servant will be responsible for effective monitoring of approvals, facilitate project implementation and handle grievances. Within the SWS, there should be a committee for monitoring and grievances, that will meet at least once a month and decision taken in these meetings shall be binding upon all ministries and departments.

The committee can also decide on issues that might create bottlenecks in the smooth process of industrialization and reviewing of policies from time to time and make its recommendations to the Economic Development Board (EDB).

Infrastructure Development Initiative.
Infrastructure Development Fund

The Federal Government recognizes the partnership role of the private sector and international development agencies in the developmental process. However, it should follow a focused approach for attracting private sector investment into infrastructure.

The Federal Government can set up the Niger Delta Infrastructure Fund (NDIF) that will act as the catalyst for mobilizing and channeling private resources, as well as that of development agencies, into infrastructure development in the region. This fund will be under the auspices of the FGN. Each State Economic Development Board can apply for funding of their various State projects from the central pool, the NDIF. In summary the NDIF will be constituted out of resources raised by the FGN and will have both institutional and private sector participation.

The fund shall be professionally managed with advisory assistance from an independent professional body to be engaged for the purpose. Individual viable infrastructure projects will be considered as referred to it by State EDBs on user charge basis for funding through Special Purpose Vehicles (SPVs). More details are provided in a later chapter in this book.

Industrial Estates.

Industrialisation in the Niger Delta must commence with the establishment of strategically located modern industrial estates throughout the region. . The industrial estates to be developed can be categorized according to the level of infrastructure and shall be priced accordingly.

High Intensity Infrastructure Industrial Estates:

These estates shall have constructed sheds, industrial plots, internal roads and parking facilities, water supply, sewage and storm water disposal systems, internal electrification, telecom facilities, solid waste disposal system, recreation centers and parks, banks, Post Offices and health facilities.

Medium Intensity Infrastructure Industrial Estates:

These estates shall have industrial plots, internal roads, water supply, open drainage system and power supply at external source.

Low Intensity Infrastructure Industrial Estates:

These estates will be built primarily for large-scale units and will be provides only approach roads and power supply at external source. The price of land in such estates will be kept very low and the units will be permitted to develop their own internal services.

In the medium and low infrastructure estates, plots will be carved out after following prudent principles of estate planning and need-based augmentation of infrastructure will be considered as the estates get populated. The development of Industrial Model Townships, Growth Centres, Technology Parks and Integrated Infrastructure Development Centres with modern facilities should be given special attention.

The project owners will develop suitable mechanisms for upkeep and regular maintenance of the estates.

Industrial Land Identification Task Force:

The purpose of setting up the above task force is to identify areas in each State that shall be notified as industrial zones to facilitate the establishment of industrial units. Such areas will be planned out for providing infrastructure at a future date either by the State Government or by private initiative or by PPPs. The procedure for change of land use (CLU) wherever required should be simplified and decentralized with a time-bound system of deemed approvals. The permission of CLU in designated industrial zones should be given by a Director of Industries with a provision for appeal to the Commissioner Industries in the State.

Power:

Power generation that was hitherto being done only by the Public sector has been thrown open to the private sector. The various State Governments should encourage the setting up of Liquid Fuel based small captive power stations of 25 MW each by the private sector at different load centers throughout the Niger Delta. State Governments in the region can also enter into new joint venture arrangements with the private sector to include coal based thermal projects, Gas-based thermal projects, hydrocarbon-based mini combined cycle power units and fuel cells for remote areas.

Power Facilitation:

The following steps should be taken to improve the interface between Industry in the Niger Delta and power distribution agencies:

- The State Governments in the region should facilitate acquisition of Distribution and Retail supply of electricity permits to specific industrial estates. These permits or licenses should be treated at par with other distribution licensees and allow the availability of bulk power at the interface point.

- Government should expedite the issuance of letters of consent for installation of captive power plants in the industrial estates as and when applied for.

- Necessary amendments in Electricity legislation should be made so that no permission for installation of captive generating units is required below a certain size, and mere intimation is sufficient.

- The Power Distribution Companies shall maintain separate seniority for release of electricity connections in industrial estates where the infrastructure is laid by the developers.

- The Government shall endeavor that all industrial units on rural feeder are shifted to urban/industrial feeders in a phased manner.

- The Power Distribution Companies shall provide temporary electricity connection for the construction period and thereafter for trial production for a period of six months for which condition of monthly minimum charges shall not be applicable.

- The process of granting permission for extension of load shall be simplified.

- Special priority for release of electricity connection to 100% Export Oriented Units (EOUs), IT Industries and FDI projects shall be given.

- A time-bound schedule for release of electricity connections shall be ensured.

Human Capital Development

Keeping in view existing and future potential of industrial development in the region, a long-term human resources development plan should be prepared in consultation with the organized private sector in the region. The industrial units that will be coming to the region will be asked to project their area specific various skills requirements They can be involved in finalizing the curriculum in Industrial training institutes, polytechnics and other institutions so that suitable orientation programmes can be organized for the local youths.

Thrust Areas.

The following thrust areas have been identified as areas to promote industrial development in the region

- Local content equipment and materials for supply to the Oil & Gas Industry.

- Agro-based and food processing industries

- Information Technology and Telecommunications.

- New gas-based industries (methanol plants, gas-to-liquids projects, fertilizer plants).

- Alternative Clean Fuels production (including Bio-fuels).

- Export oriented Units.

- Automotive components and light & medium engineering.

Establishment of the Niger Delta Industrial Development Corporation (ND-IDC).

As a matter of urgency, the State Governments in the Niger Delta region of Nigeria should set in motion the necessary machinery for the establishment of the Niger Delta Industrial Development Corporation (ND-IDC).

They can be can be assisted in this process by involving the Niger Delta Development Commission (NDDC).

Mission of ND-IDC:

The proposed ND-IDC can target a total asset base (not capital base) of about N60 billion over a period of 24 months. In other words, it can aim to provide loans and other financial facilities of about N60 billion to the manufacturing sector in the core Niger Delta States over a period of 24 months.

It can take equity positions in the ventures it will help to finance and can subsequently sell these interests to the private sector and Niger Delta citizens through the Stock Exchange.

ND-IDC will be mandated to raise long-term funds from the Bond Market for productive investment in manufacturing companies in the Niger Delta.

The ND-IDC can provide low-cost loans to increase export production and to finance exports. It can also offer low-cost loans to small businesses in the region (i.e. those with assets of less than N10 million). In general except in special cases, the ND-IDC will not participate in the management of the firms it finances.

Ownership Structure:

Proposed Ownership Structure of ND-IDC at least initially can be as follows: Niger Delta State Governments, NDDC, Private Sector, Private Placement of Shares (IPO) to Niger Delta State Citizens.

Mode Of Incorporation:

Because of tax advantages ND-IDC can be incorporated in Mauritius (a tax haven) as a holding company that will operate and have a registered office in London and the main operating subsidiary in the Niger Delta region of Nigeria. There can be branch offices in key cities in each of the stake-holder States.

Why Mauritius?:

Mauritius is not only a tax haven but is fast becoming a major offshore financial centre. The Island of Mauritius is strategically located in the Indian Ocean close to Southern Africa and well-placed to play a complementary role to that of the major financial centres in: Asia, Europe and America.

It can lay claim to providing excellent conditions: political stability; a growing and rapidly diversified economy; a strategic location and time zone; a concessionary tax regime; excellent infrastructure and support services and a well educated bi-lingual (English and French) work force. Mauritius will be the hub of DS-IDC efforts to raise foreign capital.

Essentially and specifically, this new organisation will be designed, structured and created to support the ambition and strategic objectives of the New Delta State Industrial Development Policy and its Development plans.

Proposed Location:

It is proposed that the Headquarters of ND-IDC will be Port Harcourt, Rivers State with the zonal offices in each of the stakeholder States. For registration purposes their will be a London office from inception and later a representative office in New York.

T H R I T E E N

Building a New Technically Skilled Workforce

The Niger Delta, with the assistance of the private sector must endeavour in to triple its pool of technically trained manpower over the next 15 years and this must occur alongside the development of more efficient economic structures as a result of the general move towards the markets. One of the quickest ways to achieve this is through the establishment of a regional Open University to serve the human capital development requirements of the Niger Delta.

The proposed distance education institute of learning should be established to solve the urgent problem of inadequate higher education resources in the Niger Delta in particular, especially in the fields of science and engineering. The institute will seek to bridge the digital divide and knowledge gap between the Niger Delta and the rest of the world using the power of modern information and communication technologies to provide increase access to the quality education needed to build capacity and support economic growth and sustainable development. The institute can be structured as a private, public sector partnership (PPP).

Three-way partnerships (tri-sector partnerships) involving companies, government authorities and civil society organisations can be a more effective means of reducing social risks and promoting community development. Experience working in different countries and at different stages of project development the performance of these 'tri-sector partnerships' has been systematically tested as providing greater value in terms of both business benefit and educational development impact.

There will be 3 learning centres in each of the 9 States that comprise the Niger Delta, with one centre in the chosen administrative headquarters to make 28 centres altogether in the region.

The aims and objectives of the proposed institute will thus be:

- To provide alternative centres of learning outside the conventional campus system for people of all ages; for those unable to obtain admission in the conventional Universities as well as for working professionals
- To use an innovative format, that is ICT based (via the Internet) and specifically designed to provide valuable real world education in the most convenient and efficient way possible
- To offer degrees in several areas in Science & Technology including Petroleum Technology and the Geosciences, Information Technology, Engineering the Applied Sciences, Management Science, etc.

Justification

1. Open Distance Learning (ODL) is growing in Africa and is now being recognised at both the national and institutional levels as a panacea for the democratisation of education.

2. It is fast becoming the tool for mass instruction for developing countries that are unable to reach everyone within the formal schooling system.

3. There is a general realisation that face to face classroom bound systems cannot carry the desires of government towards life-long learning and the elimination of illiteracy.

4. That indeed modern technology has eliminated the need for education within walls as it converts the world into a global village

5. Today even with 57 universities, Nigeria cannot cope with the over 1.2 million applicants to higher institutions Together they can only cater for about 25 % of all applicants. So the need to establish the open education system became more glaring

6. Historically, the provision of higher education all over the world was elitist, exclusive and limited to the rich or influential in society. The general development role of education and the need for all to be educated was not the primary concern. Today governments have realised that inclusiveness, which is to carry along the generality of its population, cannot occur unless the populace are educated.

7. No country in the world has been able to educate everyone needing it through the classroom bound face-to-face system. Today we now talk of border-less or wall-less classrooms, and affordable education for all through the use of ODL.

Some Facts

ODL has been growing very fast in Asia. Ten out of the 12 Mega Universities of the world (a University with over 100,000 students) are in Asia. The Indira Gandhi National Open University, India now has more than 1.2 million students, China more than 5 million

However while it is recognised that the Niger Delta can shorten its journey towards starting ODL by importing content from elsewhere, they must be aware that wholesale importation without adopting or adapting it to suit specific environments may be counterproductive.

At the proposed Open University it will be possible for people of all ages to earn accredited University degree while maintaining their career and personal life. It will also enable students who failed to get admission into conventional Universities to obtain a degree of their choice. Classes will be offered at times and places that meet with the convenience of the students.

The first batch of students will receive online coaching at 28 learning centres located at strategic locations throughout the Niger Delta. The curriculum will be continually updated to provide the skills and knowledge that are in highest demand.

At the proposed Open University (OU) it will also be possible to earn a degree via the Internet at home in the evenings, or at work during ones' spare time or on weekends because of the flexibility offered by the Internet.

The academic programme will also a post-graduate petroleum studies courses leading to a Masters or PHD degree to be funded by Multinational Oil Companies operating in Nigeria and other private investors, the NNPC, international donor agencies, foundations and other stake-holders. It will have affiliations with other international institutions that specialise in petroleum technology. It will be an excellent example of what can be accomplished through Public-Private sector Partnerships (PPPs) if successfully implemented.

In addition the OU will be committed to delivering focused, timely and value-added post-graduate studies for the staff of oil companies, whom on receiving their postgraduate degrees, can then be utilised to sustain a company competitive advantage.

The Oil & Gas Industry in the Niger Delta as well as the NNPC will benefit in terms of having access to a ready pool of highly trained and skilled professionals with up-to-date skills and knowledge of the industry. One of the factors that have led to the impressive growth of the Malaysian Oil & Gas Industry is the huge investment in education.

Adult Education Programs

The OU will provide adult continuing education, developed to meet the needs of adults by offering non-resident degree and other programs. The programs will be organized and developed using as a starting point the "off-campus" external degree concept tried and tested by the various "universities without walls".

The adult students will decides how long they take to complete their course work. While studying, a teacher will be available to offer help, answer questions, and guide them through the learning process, but students are free to choose their own study pace.

As an institution of distance learning, OU will to assemble a faculty of recognized experts. OU faculty will consist of university professors, business executives, consultants, and other specialists. Though these instructors will vary in subject matter expertise, they will share important attributes. All of them will be active practicing what they teach, and they will be skilled in imparting their knowledge through distance learning channels and technology.

OU will be committed to providing equal opportunities for study. SCOU educational programs will be acknowledged as the foremost, with an excellent academic reputation. OU staff and lecturers will be very committed to quality in services.

The Learning Centre

Apart from home and office coaching based on modern Internet technology, the main feature of the project is the Learning Centre. Each learning Centre will be the basic unit for delivery of lectures and mentoring

Each Learning Centre can be equipped with the following:

* A satellite dish (VSAT)
* 10 computers
* Software package
* 2 Printers
* 1 Television
* 1 VCR
* 1 Fax Machine
* 2 Telephone Lines
* 1 Radio receiver
* 1 Satellite dish (for the VCR)
* 1 Projector and screen
* 1 Generator

Human Resources Requirements:

Three types of training will be necessary.

* Training of ICT trainers: whose task will be to train the students.
* Teachers who will use ICTs to support their work
* Teachers who will be e-literate: these are important, as they will play an important and influential role in making the school goers to develop an interest in ICT.

Each Centre should have at least two ICT trainers/teachers dedicated to the students.

Human Resource Needs For The Operation Of Equipment:

There will be a need to have people whose task is to take care of the operation of the module

* Technicians for the transmission equipment
* Technicians for the user equipment
* Software

A set of technicians will be allocated to a group of schools in the same neighbourhood. Preferably these should be from around the community in which the school is found.

Awareness Campaign:
An e-awareness campaign introducing teachers and communities to ICTs will be organised in a planned and co-ordinated manner.

Curriculum Development:
Experts in the education and ICT arena that has a meaningful African input can commonly develop the curriculum that can be used eventually.

Infrastructure & Equipment Requirements:
Satellite Capacity:
Provision of satellite capacity to cover the centres to be raised with satellite operators so that they are partners.

Also once the location for the centres have been identified, it will be necessary to have a co-ordinated plan of building the necessary infrastructure

Proposed Disciplines
BSC Courses:

- Petroleum Engineering
- Chemical Engineering
- Mechanical Engineering
- Electrical/Electronics Engineering
- Civil Engineering
- Geology/Geophysics
- Accountancy
- Environmental Science
- Computer Science
- Economics
- Agricultural Science

MSC and PHD Courses:

- Energy Finance
- Petroleum Geo-science
- Petroleum Geology
- Petroleum Engineering
- Petroleum Economics

The OU can provide a comprehensive range of training and development services to support the Petroleum and Energy industries.

Recognising the functional and strategic role of energy in economic development and the need to build capacity in this sector in the region to ensure its optimal application. the, OU can form partnerships with local and international organisations to offer a broad range of specialised knowledge and skills, education and training.

At the same time the OU will strive to fill the gap in terms of skills shortages in the region and to empower previously disadvantaged communities to participate fully and effectively in the energy sector through accredited, tailored, quality programmes and courses.

The vision is to provide relevant technical training and development opportunities by creating a center of specific training excellence in the Niger Delta. The OU will strive to be the preferred service provider for all learning requirements in the Petroleum and Energy industry, thereby making a contribution to the economic growth of the country and the development of its peoples.

The OU, through our International linkages and local capability, will be able to provide petroleum and energy industry specific solutions to our clients. A strong technology focus is a key output, with the development of technical and technological skills, as key focus areas. The OU will build a unique ability to guarantee added value to learners and stakeholders through effective programmes in both up and downstream sectors, and subsequent linkage to leadership and management development, will make OU the service provider of choice. Integration of learning through nationally and internationally accredited programmes, leading towards a 'learning-to-learn' culture within the industry

Financial Resources

Financing for this project can derive from the following sources:

* The Education Trust Fund and International Foundations.
* Equity contributions/donations from Multinational Oil companies operating in the Niger Delta region.
* International Donor Agencies.
* Local Banks/Development Agencies.
* NNPC/NDDC
* Private /Institutional Investors.
* Private Foundations.

F O U R T E E N

SME & Micro-Finance Initiatives

Strictly speaking, micro finance denotes the provision of short-term operating credits to a closely defined target group of small and micro enterprises for production purposes or loans to the same group for trade.

To a limited extent finance is also provided for investment.

The target group is also given access to services in the formal sector via micro finance institutions (MFIs). This includes promoting savings facilities and insurance as well as personal loans for housing and education.

Micro enterprises are important because in the Niger Delta they account for over 90% of the economic units of the private sector and employ over 60% of the local labour force. Since small and micro enterprises usually employ particularly labour intensive production methods, substantial employment effects can be achieved with small loans.

In well-managed programs by specialists, these instruments can be put to successful use with impressively low loan loss ratios, well under 5%. Crucial for the borrower is not the interest rate, but fast un-bureaucratic and ongoing access to external funds that commercial banks do not usually offer this target group.

Promoting micro finance in development policy only make sense if the following rules are strictly observed:

* Ensure the profitability of micro finance institutions through professional management.

* Qualified capacity building on all sides.

* Commercial terms for the final borrower.

* Sustainability of services that can only be assured if the MFI does not rely on subsidised refinancing funds.

Micro-finance provides the most effective means of supporting rural entrepreneurs in the Niger Delta, by promoting micro lending and micro-savings

Micro-finance Institutions encourage and promote self-employment by placing start-up funds for income-generating activities within the reach of micro-entrepreneurs.

Specialised technical service providers (TSPs), can participate in MFI development programmes of UNCDF (United Nations Capital Development Fund), the World Bank/IFC, the ADB and other donor/ development agencies, Multinational Oil Companies, requiring the services of local NGOs that have international and global outreach. TSPs can thus offer the following services:

* Identification and screening of appropriate participating MFIs in the region

* Provide training and technical assistance to participating organisations in all aspects of micro-finance best practice

* Ensure recipient organisations have appropriate reliable and transparent reporting systems

* Assist recipients in monitoring, collecting and preventing delinquency loans

* Encourage and promote the expansion of private Micro-finance Banks as important sources of micro-finance for both rural and urban communities, while strengthening the capacity of existing ones

* Promote the establishment of new micro-finance initiatives such as Credit Bureaus, credit guarantee schemes and the use of smart cards.

* Promote the use of private leasing as alternative funding for acquiring essential income generating assets.

Longer-term funding for larger SME units:

It is a general observation that there is a critical lack of finance (especially long term investment funding) for SMEs in the Niger Delta. Some options to address these needs are as follows:

* The establishment of SME private equity funds (venture capital fund) to provide longer term financing coupled with hands-on-management and marketing assistance. The fund would actively seek out good business investments with the following characteristics:

* Operating new productive enterprises, and re structured entities.

* Potential for long-term commercial viability and growth, including significant job creation within a 2-3year period.

* Management with a willingness and interest in actively using market-oriented management practices.

The fund would provide equity financing with "hands-on" management of the enterprises, and would assume major responsibility for advising on day-to-day management and marketing.

An approach of deep mentoring assistance would be the operating mode. In the process managers of the beneficiary enterprises would receive practical experience in operating companies in market- oriented ways.

While this initiative primarily contemplates equity investments, the fund could also provide term loans as appropriate. Short-term working capital would be leveraged through local banks. It is expected that the investments would be in the US$50,000-300,000 range.

The funds exit strategy would most likely come through management buy-outs, but could also be by sale of shares to private investors or through public offerings in countries with developed capital markets. Also the fund might consider a compensation scheme based on performance in generating superior sales and profit margins.

The fund managers would determine the technical assistance needed by individual businesses. These fund managers would be free to access technical support provided through donor agencies.

Comparable funds are already operating successfully in several central and east European countries, and similar funds have recently been set up in Russia. There is thus ample precedence for this intervention as a successful method of assisting SME growth. The maximum number of beneficiary companies normally average not more than 40 per fund.

It should also be noted that a commercial venture capital fund, the Caucasus Fund LLC (for the Caucasus region), has been established and has already reached a capitalization of US$92 million. This includes an approximately 2:1 funding from the US Overseas Private Investment Corporation (OPIC).

The US Trade Development Agency (TDA) has been approached to fund feasibility studies for projects it is interested in funding. Similar Funds can be organised for the Niger Delta.

Another SME initiative, which can be replicated in Nigeria, and focussed on the Niger Delta, is the hands-on lending programs being conducted by Shore bank Advisory Services (SAS). At present SAS is lending (on-lending) IFC and USAID funds through TBC Bank (though other banks are also eligible for these partnering programs) and is effectively transferring approval and monitoring procedures, approval forms, screening procedures, and financial analysis procedures

All loans are dollar denominated, thus the borrowers bear all exchange rate risk. The IFC funded program includes loans ranging from US$10,000-250,000 (although the largest loan to date has been US$130,000) for businesses that are typically 2 years old, generally have fixed assets worth about US$100,000 and are production or service oriented.

F I F T E E N

A New Green Revolution
for the Niger Delta

Over concentration of government on Oil & Gas policy at the expense of Agriculture may have long term negative consequences for food security in Nigeria as well for the revenue accruable from the export of agricultural commodities such as rubber, palm produce that grows particularly well in the Niger Delta.

In fact, in the case of palm oil, unless Nigerian production of palm oil can be increased during the next 5-10 years, the country faces the prospect of becoming a major importer of this valuable food commodity, which forms and essential element in the peoples diet.

The development goal is:" to help start a new sustainable and equitable green revolution in the Niger Delta, one that dramatically increases the productivity of small-holder farmers, thereby improving the livelihoods and living conditions of rural communities, focusing on the six Niger Delta States. It will in addition encourage large-scale mechanised crops production using modern technological methods for increased yields. Resources will come from the Federal Government, State Governments, and the Multinational oil companies, UNDP, IFAD, WORLD BANK, ADB, FAO, Private Foundations, and Corporate Grant-Makers.

The key objectives are:
*   Empower poor rural women and men to critically analyse their constraints, opportunities, and support requirements, and to increasingly manage their own development.

* Support institutionalisation of policies and processes, create awareness and develop the capacity of the public and private sector service providers to become more relevant and responsive to the rural poor in the Niger Delta.

* Support balanced sustainable social, agricultural and economic development interventions for appropriate women and male groups (including youths) and individuals

The approach will be to expand such interventions to reach the maximum number of rural communities in the shortest possible time.

Agro-ecological conditions stand out as among the more intractable problems facing the Niger Delta, at least in the short to medium term. There is a need for improved crops varieties that address crop diseases, poor soils and other environmental problems. There are agricultural technology solutions to these problems. In fact, biotechnology offers the potential to provide the following:

- Improvement in crops that are relevant to Nigeria (e.g cassava, rice, palm oil fruits, sugar cane and maize.

- Solutions that are important to the Niger Delta such as disease resistant plants that can also thrive in poor soil conditions, as well as plants with improve nutritional value.

- Beneficiation of agricultural waste.

- Fermentation industries to add value to raw materials.

Furthermore, the Niger Delta's biodiversity includes plants, animals, microorganisms and marine species. There is huge potential to extract value from the biodiversity, for the development of:

- Novel drugs and pharmaceuticals.

- Natural plant derived medicines.

- Food products derive from indigenous produce.

- Technologies for adding value to indigenous fibre resources.

- New enzymes and metabolic pathways.

However the investment response from both the private and public sector could be better. What is needed is massive investment in R & D engagement from the oil companies and both institutional and rural physical infrastructure.

Role of Private Sector Investment.

Any private sector ambition should be to establish extensive interests in agri-business in the Niger-Delta region, but the over-riding objective should be to move, as fast as practical, into downstream processing of those primary agricultural products which have export markets, and where significant added value can be created.

Nucleus estates will be created depending on availability of adequate hectares of arable land, but also, where possible, the policy will be to encourage and support out-growers, both large and small to produce crops sufficient to supply various factories and processing plants that will be established.

With the active participation of the private sector, agriculture offers a sure route for defeating hunger, malnutrition and poverty in Africa through well co-ordinated programmes to boost food production and increase the nutrient value of what is produced. Tropical Agronomics is being set up to achieve this objective utilising the advantages modern technology has to offer.

Agriculture is still the largest foreign exchange earner in most African countries and the largest contributor to employment. The farming sector also provides a major source of income for a large part of the population. Crops as raw material source are being processed locally;

This increased local value added content generates additional employment and earning opportunities.

It is now recognised that the agricultural sector can be successfully developed by private sector investment. Private involvement means privately financed food security. This makes for additional cash income in rural areas and it eases the government's subsidy burden. In Africa, agricultural production is mainly the domain of small and medium-sized enterprises and is largely a subsistence sector.

The private sector should engage in large-scale production focusing on those crops amenable to greater yields utilising modern scientific techniques and nutrient additives to significantly boost both quantity and quality for export as well as for domestic consumption.

The company will integrate forward into processing to add value to farm products and obtaining higher export revenue by competing in international markets. It will establish an effective downstream trade and distribution system, particularly with a view to supplying the urban population and meeting a larger more sophisticated demand for its local production.

Emphasis on quality of products for export and superior packaging will contribute to the building of confidence in our preferred brands in international markets and earn customer loyalty and thus stable hard currency income.

Local co-operatives and small/micro-enterprise groups can also be integrated into our supply chain system that guarantees sales outlets for small holders.

Natural cropping conditions in the areas we have selected will enable several harvests a year and guarantee a high yield particularly when modern scientific techniques are employed that enhances yields and products disease resistant crop varieties.

There should also be a plan to have a well- trained work force of over 1500,over a period of 5-6 years and build a housing infrastructure to raise their living standards.

- Improvement to crops that are relevant to Africa (e.g palm oil, rubber, rice, soybean, maize, pine-apples, sugar cane & beet, bananas etc.)
- Plants with improved nutritional value, disease resistant plants, drought tolerant plants etc.

Local entrepreneurs should seek foreign partnerships with organizations offering both technology and funding that will enable us succeed in meeting the above objectives.

Gas-based Fertiliser Production.

With its abundant gas resources, the Niger Delta region is an ideal location for the establishment of fertiliser production plants especially by the private sector or through joint-venture arrangements with the Nigeria National Petroleum Corporation (NNPC) i.e. a Private Public sector Partnership (PPP) and local stakeholders. Nigeria that is one of the largest gas flarers in the world still flares about 60% of its gas production, leading to loss of revenue and environmental contamination.

This flared gas could be channelled into the production of fertiliser. Gas constitutes almost 90% of fertiliser production. The only manufacturing unit that exists in the region (the former government owned NAFCON) has been privatised and is now called NOTORE Chemical Industries. The current owners mainly local stakeholders, have ambitious plans for expansion utilising the regions gas resources. Other private sector operators should be encouraged to follow their example by setting up new plants that will make the country self-sufficient in fertiliser that will drive the new green revolution, create employment and wealth for citizens in the region and reduce the negative environmental impact of gas flaring. The oil & gas operators should be given all necessary incentives to participate in a massive fertiliser production programme in the region.

S I X T E E N

Exploring the Buisness Potential for Aquaculture Development in the Niger Delta

This chapter explores the possibility of breaking new ground, away from the comparatively commonplace fish farming activity into the very uncommon activity of shrimp/prawn aquaculture in the Niger Delta.

Studies have indicated the environmental suitability of the Niger Delta for shrimp farming, with an emphasis on the black tiger shrimp *Penaeus monodon*, based largely on research and practical experiences gathered from the global consortium program on "Shrimp Farming andEnvironment" and other sources

General Information of Shrimp/Prawn Farming.

Modern shrimp farming, the production of marine shrimp in impoundments, ponds, raceways and tanks, got started in the early 1970s, and, today, over fifty countries have shrimp farms. In the Eastern Hemisphere, Thailand, Vietnam, Indonesia, India and China are the leaders, and Malaysia, Taiwan, Bangladesh, Sri Lanka, The Philippines, Australia and Myanmar (Burma) have large industries. In the Western Hemisphere, Mexico, Belize, Ecuador and Brazil are the leading producers, and there are shrimp farms in Honduras, Panama, Colombia, Guatemala, Venezuela, Nicaragua and Peru. The shrimp importing nations--the United States, Western Europe and Japan-- specialize in high-tech "intensive" shrimp farming (more about this below), but, thus far, their production has been insignificant. In the Middle East, Saudi Arabia and Iran produce the most farmed shrimp.

Shrimp farms use a one-phase or two-phase production cycle. With the two-phase cycle, they stock juvenile shrimp from hatcheries in nursery ponds and then, several weeks later, transfer them to growout ponds. With the one-phase cycle, the nursery ponds are eliminated, and the shrimp are stocked directly into growout ponds, after having spent a short period in acclimation tanks (more below). Farms usually produce two crops a year, although farms within 10 degrees of the equator sometimes get three crops a year.

Hatchery

Hatcheries sell two products: nauplii (tiny, newly hatched, first stage larvae) and postlarvae (which have passed through three larval stages). Nauplii are sold to specialized hatcheries that grow them to the postlarval stage. The Hatchery Cycle: Whether gravid (ready-to-spawn) shrimp are captured in the wild or matured in the hatchery, they invariably spawn at night, but with photoperiod manipulation, they can be induced to spawn at any time. Depending on a number of variables (temperature, species, size, wild/captive and number of times previously spawned), they produce between 50,000 and 1,000,000 eggs.

After one day, the eggs hatch into nauplii, the first larval stage. Nauplii, looking more like tiny aquatic spiders than shrimp, feed on their egg-yoke reserves for a couple of days, and then metamorphose into zoeae, the second larval stage, which have feathery appendages and elongated bodies but few adult shrimp characteristics.

Zoeae feed on algae and a variety of formulated feeds for three to five days and then metamorphose into myses, the third and final larval stage. Myses have many of the characteristics of adult shrimp, like segmented bodies, eyestalks and shrimp-like tails. They feed on algae, formulated feeds and zooplankton. This stage lasts another three or four days, and then the myses metamorphose into postlarvae. Postlarvae look like adult shrimp and feed on zooplankton, detritus and commercial feeds.

Farmers refer to postlarvae as "PLs", and as each day passes, the stages are numbered PL-1, PL-2, and so on. When their gills become branched (PL-13 to PL-17), they can be moved to the farm. From hatching, it takes about 25 days to produce a PL-15.

Small-Scale Hatcheries: Hatcheries come in three sizes: small, medium and large. A family group on a small plot of land usually operates small-scale hatcheries. Called "mom-and-pop" or "backyard" hatcheries, they adopt a green thumb, non-technical approach. Their chief advantages: low construction and operating costs and the ability to open and close, depending on the season and market factors.

They utilize small tanks (less than 10 tons) and concentrate on just one phase of production, like nauplii or postlarvae production. They often use low densities and untreated water. Diseases, the weather and water quality problems often knock them out of production, but they can quickly disinfect and restart operations. Survival of the developing larvae in small-scale hatcheries ranges from zero to over 90%, depending on a wide range of variables, like stocking densities, temperature and the experience of the hatchery operator.

Small-scale hatcheries have achieved great success in Southeast Asia, particularly in Thailand, Taiwan, Indonesia, the Philippines and southern China. In Thailand, which has thousands of backyard hatcheries, the industry is segmented into suppliers of nauplii, postlarvae, phytoplankton, equipment, feeds and chemicals.

Medium-Scale Hatcheries: Most medium-scale hatcheries are based on a design developed in Japan and popularized by the Taiwanese. Called "Japanese/Taiwanese", "eastern" or "green water" hatcheries, they use large tanks, low stocking densities, low water exchange and encourage an ecosystem to bloom within the tank.

This bloom feeds the developing shrimp. In some cases, various nutrients and bacteria are added to the tanks that discourage the growth of "bad" bacteria and encourage the growth of "good" bacteria (probiotics). This ecosystem approach is supposed to produce stronger postlarvae due to its closer approximation of natural conditions and the absence of therapeutics. Survival, from stocking to harvested postlarvae, is usually 40%, or less.

Large-Scale Hatcheries: These are multimillion-dollar, high-tech facilities that produce large quantities of seed stock in a controlled environment. Originally developed at the Galveston Laboratory of the United States National Marine Fisheries Service, they are referred to as "Galveston", "western" or "clear water" hatcheries.

Requiring highly paid technicians and scientists, they utilize big tanks (15 to 30 tons), filtered water, high densities, and high rates of water exchange, allowing them to take advantage of the economy of scale by producing seed stock throughout the year. They grow algae and brine shrimp and feed them to the developing shrimp. High survivals, up to 50%, are common with these systems, though in practice survivals range from zero to 80%.

In the Western Hemisphere, big hatcheries are the established trend, but large-scale hatcheries can also be found in all the major shrimp farming countries.

Many large-scale hatcheries maintain captive broodstock in "maturation facilities", which require expensive live feeds like bloodworms, squid, bivalves and other crustaceans (adult *Artemia* and krill). Dry formulated feeds are not as popular because they don't work on a 100% replacement basis.

Since *Penaeus vannamei* (the most popular species in the Western Hemisphere) is easier to work with than *P. monodon* (the most popular species in the Eastern Hemisphere), captive breeding is more common in the west than the east. Most breeding facilities recirculate the water in the broodstock tanks, creating a closed system where water quality variables can be controlled and external factors limited.

Hatchery Feeds: Hatcheries utilize a combination of live feeds, such as micro algae and brine shrimp nauplii (*Artemia*), with one or a number of prepared diets, either purchased commercially or prepared at the hatchery. The principal algal species employed is *Chaetoceros muelleri*. Again, dry formulated feeds are popular, but they don't work on a 100% replacement basis.

Eighty percent of the hatcheries in the Western Hemisphere use some artificial broodstock diets. In 15% of the hatcheries, artificial diets represented more than 25% of the total feeding regime. Hatcheries used Breed S (INVE Aquaculture NV, Belgium), Higashimaru (Higashimaru Co., Japan), MadMac MS (Aquafauna Biomarine, Inc., USA), Nippai (Japan), Rangen (Rangen, Inc., USA) and Zeigler (Zeigler Bros., Inc., USA).

Hatcheries feed *Artemia* nauplii nutrients and medicines and then feed those nauplii to larval farm-raised shrimp--and the nutrients and medicines are passed on to the shrimp. The *Artemia* naups can be spiked with bactericides to reduce the bacterial loads during hatching and holding.

Freeze-drying (lyophilization) removes water from frozen products without having to thaw them first. The process of freeze-drying is considered to be the most conservative and safe method for drying and preserving fragile nutrients.

Hatchery Trends: In the Western Hemisphere, hatcheries are usually very large and often associated with big farms. They frequently supply nauplii to smaller hatcheries in other regions and other countries. The smaller hatcheries raise the nauplii to postlarvae, which are sold to farms for stocking in nursery or growout ponds. Many of the large centralized hatcheries breed shrimp for special characteristics, like rapid growth and disease resistance.

In the Eastern Hemisphere, small and medium-scale hatcheries continue to produce most of the seed stock.

Worldwide, the once clear distinction between Japanese/Taiwanese-style and Galveston-style hatcheries is increasingly blurred, as a large number of hybrid operations, borrowing the best from both, are adapted to local conditions and experience. The advent of the backyard hatchery has further blurred the distinction. Success has not been the exclusive domain of any one style, and it is becoming more and more obvious that hatcheries must be adapted to local conditions.

NURSERY

The nursery phase of shrimp farming, when postlarvae are cultured at high densities in small earthen ponds, tanks and raceways, or even in inclosures within the growout ponds, occurs between the hatchery and growout phases. Since hatchery-produced and wild-caught postlarvae can be stocked directly into growout ponds, the nursery phase is not always necessary.

Farmers stock postlarvae in nursery ponds (0.5 to 5.0 hectares) at densities of 150 to 200 per square meter and feed a crumbled diet several times a day. Protein levels in these feeds range from 30 to 45%. The nursery phase should not exceed 25 days.

Proponents of nurseries argue that they improve inventory, predator and competition control; increase size uniformity at final harvest; better utilize farm infrastructure; permit more crops per year; improve risk management; produce stronger postlarvae; and decrease feed waste. Because low salinity levels can be lethal to postlarvae, nurseries also provide a halfway house where salinities can be adjusted to pond levels.

The main criticism of nursery systems is that postlarvae suffer mortalities when they are transfered to growout ponds. Spontaneous mortalities also occur in nursery ponds when animals are held beyond 25 days.

Nurseries in greenhouses find applications in temperate climates where it is important to get a jump on the growout season.

Shrimp farmers used to hold animals in nursery ponds for 30 to 60 days; now they try to move them into growout ponds in less than 30 days (at 0.5 to 0.8 grams each). This reduces stress on the animals and dramatically increases survivals in the growout ponds. Many farms in Honduras that abandoned nursery ponds have gone back to them, and the results have been surprisingly positive. They're using the old, uncovered, earthen, nursery ponds.

GROWOUT

Once a growout operation is stocked with postlarval shrimp, it takes from three to six months to produce a crop of market-sized shrimp. Northern China, the United States and Northern Mexico produce one crop per year, semi-tropical countries produce two crops per year, while farms closer to the equator have produced three crops a year, but rarely. Temperature has a lot to do with it. Shrimp like it hot, and most species prefer, but are not restricted to, brackish water.

Growout operations come in all shapes and sizes. They are classified by stocking densities (the number of seedstock per hectare) and called "extensive" (low stocking density), "semi-intensive" (medium stocking density), "intensive" (high stocking density) and "super-intensive" (highest stocking density). As densities increase, the farms get smaller, the technology gets more sophisticated, capital costs go up and production per unit of space increases dramatically.

Extensive: Extensive shrimp farming (low-density) is conducted in the tropics, in low-lying impoundments along bays and tidal rivers, often in conjunction with herbivorous fish. Impoundments range in size from a few hectares to over a hundred hectares. When local waters are known to have high densities of larval shrimp, the farmer opens the gates, impounds the wild larvae and then grows them to market size.

Fishermen also capture wild postlarvae and sell them to extensive farmers for stocking. Overall, however, stocking densities are quite low, not over 25,000 postlarvae per hectare. The tides provide a water exchange rate of from 0 to 5% per day. Shrimp feed on naturally occurring organisms, which may be encouraged with organic or chemical fertilizer. Construction and operating costs are low and so are yields. Cast-nets and bamboo traps produce harvests of 50 to 500 kilograms (head-on) per hectare per year.

Semi-Intensive: Conducted above the high tide line, semi-intensive farming introduces carefully laid out ponds (2 to 30 hectares), feeding and pumping. Pumps exchange from 0% to 25% of the water a day. With stocking rates ranging from 100,000 to 300,000 postlarvae per hectare, there is more competition for the natural food in the pond, so farmers augment production with shrimp feeds.

Wild or hatchery-produced postlarvae are stocked in growout ponds, which are fertilized (nitrogen, phosphorus and silicate) to encourage a natural food chain. The farmer harvests by draining the pond through a net, or by using a harvest pump. Yields range from 500 to 5,000 kilograms (head-on) per hectare per year. Farmers usually renovate their ponds once a year. If too many semi-intensive farms concentrate in a small area, they can have a negative effect on the environment.

Intensive: Intensive shrimp farming introduces small enclosures (0.1 to 1.5 hectares), high stocking densities (more than 300,000 post-larvae per hectare), around-the-clock management, heavy feeding, waste removal and aeration.

General Overview of the Coastal Environment in the Niger Delta region.

In general terms, the Niger Delta of Nigeria has many characteristics that make them environmentally suitable for shrimp farming (the countries are also on the same latitude as shrimp producing countries in Asia growing *Penaeus monodon*, and in South America, on the other side of the Atlantic, where white shrimp are cultured).

Site suitability

Within the coastal areas of the Niger Delta, there are coastal land areas that (superficially at least) are physically suitable for shrimp farming in all countries.

Extensive areas of flat land behind mangrove areas can be seen, as well as more open coastal flats where ponds might be constructed. However, detailed understanding of local conditions is essential to make decisions on suitability of different areas.

Coastal Mangroves and Other Coastal Habitats

Coastal habitats in the Niger Delta region vary considerably, ranging from rocky cliffs, and extensive sea grass prairies in the north Niger Delta to dense mangrove forests and well-developed estuaries in the south

A critical issue is therefore where shrimp farming will fit into the existing habitats and land use patterns in coastal areas of the region.

A further development issue will be to minimize impacts on mangrove forests that are extensively found through the region

Other countries have smaller areas of mangroves. This region contains mangroves of the Niger Delta, which is the single most extensive mangrove system in Africa, and third worldwide after India and Indonesia.

Mangroves throughout the region are threatened, with over-cutting for firewood for cooking, building and salt drying, having affected natural habitats. A marine protected areas strategy for the northern part of the region has been prepared that includes mangrove habitats.

Clearly, careful attention will be needed to development of shrimp farming in the region, and particularly in areas where there are significant coastal mangrove resources, to ensure that aquaculture does not impact on mangroves adding to the existing damage.

The need for protection of critical habitats is increasingly being recognized, and there is a number of Global Environment Facility (GEF) projects in the region supporting identification and building of institutions and policy frameworks for conservation and management of these critical habitats.

Ideally, shrimp aquaculture development should proceed alongside such initiatives, where suitable areas for shrimp and other forms of aquaculture are identified and properly zoned in ecologically less sensitive areas, within coastal management plans, and shrimp farming is restricted in ecologically sensitive areas.

Another feature of the coastal areas in the Niger Delta is the "mangrove swamp rice environment". According to studies by WARDA, rice is also grown on approximately 190000 ha of mangrove swamps, representing 5% of area and producing roughly 7% of the region's production.

Located on tidal estuaries close to the ocean, most mangrove swamps experience a salt-free growing period during the rainy season when freshwater floods wash the land and displace tidal flows. As a result, the rice-growing period is directly related to distance from the ocean, varying between less than four months in the nearest estuaries to more than six months in those more distant.

Soils are generally more fertile than in the other environments since they benefit from regular deposits of silt left during annual flooding. However, the soils are also characterized by high salinity and sulfate acidity. Lower rainfall during the last two decades has reduced seasonal flushing substantially, further accentuating both problems.

Some of these areas, may be well suited to shrimp farming, however, there are also risks of conflicts with rice farmers, and impacts on rice farming if shrimp culture projects are not properly planned and implemented.

As the livelihoods of many people living in coastal areas of the Niger Delta are also be dependant on mangroves and other coastal resources, it will be important to be aware of potential conflicts with other coastal habitat users when selecting sites.

Apart from the rice and mangrove areas, are so-called "tannes" or the non-exploited zones that are widely recognized as having scope for aquaculture development, making use of areas that cannot be used for agriculture.

CLIMATE

The Niger Delta region has wet and dry seasons resulting from the interaction of two migrating air masses. The first is the hot, dry tropical continental air mass of the northern high pressure system, which gives rise to the dry, dusty, Harmattan winds which blow from the Sahara over most of West Africa from November to February; the maximum southern extension of this air mass occurs in January between latitudes 5° and 7°N.

The second is the moisture-laden, tropical maritime or equatorial air mass, which produces southwest winds. The maximum northern penetration of this wet air mass is in July between latitudes 18° and 21°N.

In the Niger Delta temperatures are much more stable throughout the year, and most favourable for shrimp farming.

The Niger Delta region is also being affected by, and is highly vulnerable to climate change, that is expected to increase average temperatures and make rainfall more erratic. Aquaculture has a potentially important role to play in such conditions, as a source of food and employment, but careful integration into coastal zone management planning will be essential, even though complex.

Coastal water quality

In general, the coastal areas of the Niger Delta region have water suitable for farming of *Penaeus monodon*, and other shrimp species. Water salinity can be found within acceptable limits throughout the region, although the large delta areas are subject to wide fluctuations on salinity that has an influence on shrimp farming.

In such cases, stocking practice and timing will need to be modified to take account of periods of high water salinity fluctuation. Water temperatures are generally suited for *Penaeus monodon* shrimp farming.

Water pollution is a concern around urban developments in the region. In several parts of the region, for example rapid increases in population, industrialization and urbanization have been experienced in the last 20 years.

Population growth due to high birth rates and migration from interior provinces has resulted in a population density of 250-300 persons per square kilometre along the Atlantic coast. The fast growing cities have been unable to provide the requisite sanitation. In some cities, only 1.3 per cent of households have access to sewerage, while in others there are no facilities for wastewater treatment. These factors have contributed to significant degradation of the natural resources and biodiversity of the coastal and international waters of the Niger Delta and adjacent freshwater catchment areas.

These have an influence on water quality, and will make locations close to large cities risky and less suitable for shrimp farm development.

Shrimp hatcheries require good quality full strength seawater (28-35 ppt) that can be found along several coastal areas in the Niger Delta, and there are a number of islands in the region where hatcheries could be located with suitable uncontaminated water supplies.

Several other islands in the Niger Delta region are also physically suited for hatchery development.

Soil conditions

Lowland coastal soils in the region include several types, including acid sulphate sand sandy soils. Acid sulphate soils are common throughout the Niger Delta region of West Africa, particularly near mangrove areas. Whilst suitable soils exist, careful site selection is required to properly assess soil conditions, and site farms in areas with suitable soils.

OPERATIONAL FEATURES
Shrimp seed

A surprising finding is that *Penaeus monodon* is widely found in the wild along the coast of the Niger Delta, probably the result of an earlier introduction. This resource provides the potential for development of a *P. monodon* hatchery system based on local resources, avoiding the need to introduce shrimp from other regions, and risks of import of disease. Further assessment of the available resources would be useful to determine the extent of stocks available for shrimp farming and assist in business planning.

Feeds and feed supply

Shrimp feeds are available in the region, but most is currently being imported. Development of feeds based on local production will be sustainable, and in any event will reduce the additional costs and problems associated with customs and import procedures. Some research, or perhaps cooperation with an Asian business with experience in shrimp feed, is recommended to develop locally suitable feeds, ideally based on local resources.

Integrated farming practices

There are interesting possibilities of integration of shrimp farming with other aqua- or agriculture-based activities in the Niger Delta, enabling resources to support a broader range of livelihood opportunities.

Importance of Human Capacity

To implement better farming practices, shrimp farms also require management systems and skilled staff. The development of skilled manpower, generally absent at the present time, will therefore be a critical factor to support shrimp farm development in the Niger Delta.

How can the optimal operational solutions be developed and promoted in the Niger Delta?

Technical solutions are available for environmental management, and these could be adapted and promoted in the Niger Delta through the following:

- Development of operational guidelines relevant to the Niger Delta situation

- Training, education and awareness raising for policy makers, government and private sector managers and farm staff.

- Supporting shrimp farming projects in the Niger Delta with better potential and sharing experiences from these projects as "models" from other countries in the region.

- Promoting optimal solutions through market access arrangements to niche export markets in Europe with high environmental and social standards, thus providing a market incentive for strict operational standards, addressing the key environmental and social issues.

- Implementation supported by building of legal and policy frameworks for shrimp aquaculture. In general, the approach is practical and feasible in the Niger Delta, but capacity building and technical support are an essential requirement.

Government has an important role to play, and it is likely that only countries with reasonable government policies and political stability will be most attractive to investors.

As a follow up recommendation, operational guidelines on better management practices for shrimp farming in West Africa should be developed, as a check list for investors in shrimp farming.

Human capacity for advising developers is also weak, suggesting the need for further training and awareness building.

A set of operational guidelines can also be a useful prelude to development of legislation.

Several countries also have common problems, although there are differences. Location and siting in mangrove areas would be of concern in countries with countries with important mangrove resources for example.

Cooperation and information exchange can be promoted across countries with similar problems to assist in seeking and sharing solutions, and learning.

If aquaculture does develop in the Niger Delta, it will need significant support in terms of personnel and facilities. A new fishery and aquaculture institute can be established at the University of Port Harcourt, within the Department of Biology of the Faculty of Sciences. This can be a positive move to help develop personnel in the region, and can thus provide a basis for support aquaculture development in the Niger Delta, and service other parts of West Africa as a regional centre.

Role of Government

Adequate policy and a functional regulatory framework are necessary for sustainable shrimp farming. In many countries of West Africa, environmental management guidelines and enabling legislation, as well as adequate human and institutional capacity to implement these policies, are lacking.

Government has an important role to play in overall regulation, in partnership with the business sector, by creating the awareness and putting the legal framework in place.

Often recommendations on legislation and institutional development in aquaculture are stated in general terms, or very broad in scope. This section therefore analyses the requirements against the key environmental factors identified above, focusing attention on the key requirements.

Environmental Impact Assessment

Although environmental impact assessment is not universally applied for aquaculture development in Asia, or Latin America, it is recognized as an important tool in the environmental planning process for aquaculture, and it is recommended in the Niger Delta.

Environment impact assessment can be applied to aquaculture at different levels:

- Individual project level, where the environmental impacts of individual aquaculture projects are assessed, and mitigation methods identified and management strategies put in place
- Coastal region or sectoral level, where the environmental impacts of aquaculture plans and particularly the cumulative effects of multiple developments can be assessed.

Although Nigeria has an Environmental Protection Agency and legislation, enforcement of standards is limited. Environmental Impact Assessments are required for aquaculture development, but there are no apparent standards, which need to be met.

There are also several new initiatives supporting coastal management in West Africa, as well as regional initiatives on marine (World Bank Coastal Biodiversity Management Program (PO49513). The project has an objective of conservation, and sustainable development, with an emphasis on livelihood improvement in coastal protected areas development. These present opportunities for integrating aquaculture in the context of coastal planning, providing a source of livelihood within the framework of a coastal resources management.

Dialogue with such projects should be initiated and areas for coastal aquaculture designated within such plans.

However, it should recognize that institutional, economic, and legal capacities of most Governments in the region remain ill equipped and under funded in dealing with environmental problems. There is a need for specific guidance on EIA for aquaculture, capacity building and development of standards that can be used among the countries of the region.

To facilitate foreign investment in the aquaculture sector, strategic environmental impact assessments should be conducted, as part of the process of identification of suitable aquaculture sites for zoning and development of overall aquaculture plans.

Understanding and Consensus on Shrimp Farming in the Niger Delta

Government commitment is extremely important if shrimp aquaculture is to develop in the Niger Delta. Consensus among other stakeholders is also equally important. There are many negative perceptions of shrimp farming globally and the use of land and it will be necessary to develop good communication and local community involvement from the outset of any project development. This will be a government, institutional and entrepreneurial task.

Summary –There is a definite possibility of developing sustainable shrimp farming according to international practices in the Niger Delta.

The FAO Code of Conduct for Responsible Fisheries provides a framework (together with outcomes from the Consortium program on shrimp farming and the environment) for identifying the requirements.

127

Firstly, the biological and physical environments along the Niger Delta coast are suitable for shrimp farming. The environments are also likely to prove attractive to investors, provided underlying business risks can be resolved.

Secondly, considerable institutional and policy constraints exist for shrimp farming.

There is a need for development of a suitable legal framework and policy for shrimp farming, and to start to put in place the necessary human resources to support its sustainable development.

There are several opportunities for cooperation in shrimp farming between Asia and the Niger Delta region of West Africa, covering areas for direct investment as well as cooperation in skills development and sharing of experiences. The following is an initial set of recommendations for follow up on these topics.

1. It is recommended that publicity material be prepared; ideally short, providing an overview of the shrimp farming opportunities in the Niger Delta, for wider circulation in Asia. In Asia, the information could be circulated through various channels to entrepreneurs and Governments, some of whom (such as Thailand, Malaysia) have policies increasingly positive towards cooperation with Africa.

2. Prepare an investment guide for shrimp farming in the Niger Delta region of West Africa, that would include relevant information on comparative advantage, technical, economic, environmental and social issues and the permitting/investment processes required for entrepreneurs should they wish to invest. The guide could be prepared with inputs from Asian aquaculture entrepreneurs and expertise, and ideally in a participatory way involving inputs as far as possible from all West African countries.

3. A workshop in the Niger Delta would be useful to bring relevant people (entrepreneurs, key policy actors) together from West Africa, and Asia, to share experiences in shrimp

farming, perhaps develop the above mentioned guide, using the workshop to raise the profile of the study among key Asian and West African entrepreneurs/policy agents, identifying the key factors for successful shrimp farming and starting a process of encouraging cooperation among West African countries in shrimp farming.

4. The Organised Private Sector in partnership with Government Agencies such as the Niger Delta Development Commission (NDDC) should start to encourage regional cooperation in shrimp aquaculture, and perhaps more generally aquaculture, among countries in West Africa. Stakeholders in the Niger Delta and other West African countries should be encouraged to start shrimp farming in the easiest "cluster", but agencies should then actively encourage sharing of these experiences among all clusters. Regional collaboration in aquaculture has proved useful in Asia to provide mutual assistance. and raise skills levels through sharing of experiences and expertise, and could prove equally so in the Niger Delta region of West Africa.

5. Shrimp and other forms of aquaculture have significant potential in the Niger Delta region of West Africa. However, the skills base among entrepreneurs and government management/policy makers for management of the sector's development is poor .A south-south program of skills development in aquaculture, sharing experiences among Asia and Niger Delta through short and long term training/educational programs and technical exchanges has potential to raise skill levels in the region, as well as creating more awareness of Niger Delta aquaculture potential among interested Asian partners. NDDC might consider supporting a south-south cooperation program for skills development in aquaculture.

6. In the Niger Delta, there appear to be areas where immediate technical assistance from Asia might provide early benefits for aquaculture development. . In the Niger Delta interest in breeding and culture on local freshwater prawns might be stimulated by sharing of knowledge in prawn breeding and farming from several countries in SE Asia. An organised

program of cooperation involving movement from Asia to Niger Delta and vice-versa could be constructed to capitalize on such opportunities.

7. A weakness is the legal framework and limited understanding or implementation of international standards for shrimp farming, including notably the FAO Code of Conduct for Responsible Fisheries (CCRF). Technical support, perhaps through a regional collaborative initiative, could be provided to raise awareness of the CCRF, and require legal measures that can be taken to support its implementation.

8. In 2007, we are proposing there be 2-3 practical 2-3week training programs on shrimp farming in the Niger Delta with facilitators from Asia and participation from West Africa generally.

There are also other aquaculture research and training agencies in Asia, including ICLARM, SEAFDEC, AIT and others that could provide useful training opportunities for the Niger Delta.

SEVENTEEN

A Regional Hub for Bio-Fuels Production

In the Niger Delta, we should be focussed on creating a future that includes renewable energy. Our Nations growing dependence on fossil fuels and the impact of our consumption have never been more apparent than they are today. There is an alternative way-renewable fuels.

The Federal Government of Nigeria is introducing the use of ethanol in fuel to be spear headed by the Nigerian National Petroleum Corporation (NNPC) although it will essentially be private sector-driven.

The domestic demand alone for ethanol is 180 million litres. All ethanol is imported In Nigeria. Local suppliers today in Nigeria are satisfying none of these markets.

For Nigeria to meet its national demand for ethanol it needs about 120 small scale (5000 litres/day capacity) plants. One plant of this size will be able to provide employment for at least 400 to 500 persons along the commodity chain in one year of 300 days. At the moment Thailand is repositioning itself to compete in the international ethanol market. International interest in ethanol production is driven by demand for bio fuels.

Many countries are currently pressurised by the Kyoto Protocol to reduce consumption of fossil fuel. Since Nigeria is a fuel producing country, it can only save foreign exchange by emphasising domestic production of ethanol from sugar cane ,cassava and other cereals for the industrial pharmaceutical and beverage market in the domestic sector.

The Niger Delta with its abundance of marshland suitable for sugar cane production can take advantage of these natural endowments to become a regional hub for bio-fuels production from sugar cane and cassava. This will create employment in the region and contribute to the reduction of fuel prices.

Ethanol can help reduce global warming because less carbon dioxide is released into the atmosphere than with conventional gasoline (petrol). Because it is an environmentally friendly source of high-octane fuel, it is widely used as a blending ingredient in petrol, more than four billion gallons per year

Ethanol is a high octane anhydrous (water free) alcohol produced by fermenting converted starch with yeast. It can be produced from any biological feed-stock that contain appreciable amounts of sugar or materials that can be converted into sugar such as starch or cellulose. Sugar beets and sugar cane are examples of feedstock that contain sugar. Corn contains starch that can relatively easily be converted into sugar. A significant percentage of trees and grasses are made up of cellulose, which can also be converted to sugar, although with more difficulty than required to convert starch.

The ethanol production process starts by grinding up the feedstock so it is more easily and quickly processed in the following steps. Once ground up, the sugar is either dissolved out of the material or the starch or cellulose is converted into sugar. The sugar is then fed to microbes that use it for food, producing ethanol and carbon dioxide in the process. A final step purifies the ethanol to the desired concentration.

Ethanol is also made from a wet-milling process. Many larger ethanol producers use this process, which also yields products such as high-fructose corn sweetener

Ethanol made from cellulosed biomass materials instead of traditional feedstock (starch crops) is called *bioethanol*.

In the USA, the Clean Air Act Amendments of 1990 mandated the sale of oxygenated fuels in areas with unhealthy levels of carbon monoxide. Since that time, there has been strong demand for ethanol as an oxygenate blended with gasoline. In the United States each year, approximately 2 billion gallons are added to gasoline to increase octane and improve the emissions quality of gasoline.

Blends of at least 85% ethanol are considered alternative fuels under the Energy Policy Act of 1992 (EPAct). E85, a blend of 85% ethanol and 15% gasoline, is used in flexible fuel vehicles (FFVs) that are currently offered by most major auto manufacturers in the USA. FFVs can run on gasoline, E85, or any combination of the two and qualify as alternative fuel vehicles under EPAct regulations. of the US.

In some areas, ethanol is blended with gasoline to form an E10 blend (10% ethanol and 90% gasoline).

The ethanol produced will be primarily for export to the USA. It will be like outsourcing of ethanol production to Nigeria where there is comparative economic advantage.

The recent spike in oil prices and environmental concerns involving the use of fossil fuels has ignited interest in ethanol production. However, there are other factors involved with ethanol production. Ethanol production spills over into other areas including the use of by-products for livestock feed, adding energy crops to crop rotations and increasing demand for cassava.

The objective of this project is to extend the use of bio-ethanol produced from cassava that is abundantly grown in Nigeria as a bio-energy fuel .The ethanol will be supplied by the establishment of a bio-ethanol production plants using cassava tubers or sugar cane as feedstock

Conventional energy strategies rely on supply- focused, fossil intensive, large-scale approaches that do not address the needs of the majority of consumers. Bio-fuels are an important potential contributor to sustainable energy strategies. The US Environmental Protection Agency refused California's request for a waiver from the Clean Air Act's requirement to include oxygenate in the state's reformed gasoline This resulted in a major demand for ethanol as a substitute for MTBE.

Ethanol plants can contribute to meeting the needs of the US market and the bio-energy needs of Nigeria.

The bulk (60%) of the production can initially be exported to the US, while the remaining 40% will be reserved for domestic needs. The innovation lies in the establishment of an ethanol plant, which will be the first of its type ever built in Nigeria that uses cassava tubers as feedstock, Conventional energy fuels, such as Fuelwood, Kerosene, diesel or petrol are still the dominant fuels for this purpose, with all their negative side effects that are well known.

Bio-diesel from Palm Oil.

Biodiesel is a clean burning renewable fuel, commonly derived from natural vegetable oil. There is no reason why it cannot also be made from palm oil (or palm kernel oil) that can be produced in abundance in the Niger Delta. While it contains no petroleum, bio-diesel can be easily blended with petroleum diesel to run in all existing diesel engines.

Biodiesel is an environmentally friendly fuel source because it is made from renewable resources and thus has lower emissions compared with petroleum diesel. It is readily biodegradable and non-toxic. Since it can be made from palm oil, it production will stimulate the Niger Delta palm oil industry and create employment and wealth for the region from both export demand and local consumption. The residue that is left over during the process can be used for animal feed.

E I G H T E E N

Security & Human Rights Issues in the Niger Delta

It is broadly recognised that stability and peace are prerequisites for poverty alleviation and a successful development process and that sustainable development, when successfully pursued, reinforces human security, stability, and peace. Years of investment in development projects have been destroyed because of violent conflict, and decades of development gains can be wiped out by one major crisis.

The oil companies work in a difficult environment in Nigeria, both physically and politically. In particular, the political environment is one in which the Nigerian government has failed to ensure that the people who live in the oil-producing areas actually benefit from the oil. But the oil companies are also seen to have failed to give back anything to the delta for what they have taken out and are often a more accessible-and responsive-target for protest than the government.

Following years of employee abductions and hostage taking, repeated protests, including occupations of their facilities that close down production, the oil companies now have quite extensive programs for community development projects in the "host communities" for oil facilities, make substantial payments for allowing oil work to be carried out both to local government authorities and to other interest groups in the areas where they are working, and frequently hire youth for "surveillance contracts" in order to satisfy a demand for employment that cannot be met in this capital- rather than labor-intensive industry.

In other cases, they hand out cash payments, sometimes to legitimate representatives of the communities where they operate, as compensation for spills, for example; but often to individuals or groups who have gone into hostage-taking or oil facility occupation as a means of earning a living. But these payments, even the best intentioned, have themselves generated problems.

In addition, the cash economy created by oil undermines those trying to work for longer-term and more sustainable development initiatives. As one development expert noted: "Anything that does not deliver instant cash, people are not interested.

Shell in Nigeria has increased its spending on community development projects greatly over the last decade and has created an entire community development unit within the Shell Petroleum Development Company Ltd (SPDC) to administer this money and attempt to redirect community relations from handing out cash to proper development schemes. In 2006, SPDC spent approximately U.S. $72 million on community development in the Niger Delta. While this development spending has undoubtedly brought benefits to the delta, much of the money has not been effectively used. According to an evaluation of Shell Nigeria's development projects carried out in 2006 by outside consultants paid for by Shell, less than a third of 408 projects were considered fully successful.

In addition, SPDC paid U.S. $12.1 million in 2006 as compensation in respect of third-party claims resulting from oil spills and construction damage (none of the oil companies pay compensation if, for example, an oil spill is the result of vandalisation of pipelines or wellheads).

SPDC has also held a number of "stakeholders' workshops" on its operations, attended in the last two years by several hundred individuals from nongovernmental organizations, various levels of government, journalists, academics, and community representatives. Although the format of the workshops, in particular their large size, means that little in the way of concrete results can be expected from them, they do provide a forum at which some people who would not otherwise be able to do so have an opportunity to express frustrations and criticisms to the company, and some company employees can be exposed to and learn from those frustrations.

Shell has also taken steps to improve environmental practice, for example through the ISO 14001 certification process, and by initiatives to improve the quality of the environmental impact assessments carried out before new oil exploration or production can take place. While much remains to be done, these efforts have reportedly had some positive results.

Overall, Shell has made serious efforts to improve its performance in Nigeria but that these efforts have in too many areas yet to yield meaningful results on the ground. The effects are largely visible only to those who have access to information about Shell's operations at quite a high level.

Many Niger Delta communities are fragile though they may not be experiencing violent conflict, and the engagement of outside actors, even in seemingly unrelated sectors, is likely to have a significant impact on the way that a country's political, social, and economic tensions evolve or are resolved.

The OECD Development Assistance Committee (DAC) has identified an irrefutable link between conflict, peace, and development, and a Policy Statement and Guidelines on Conflict, Peace, and Development, issued in May 1997, clearly placed peace-building and conflict prevention on the development agenda. Sustainable development cannot be achieved without being sensitive to the tensions that divide communities.

The DAC Guidelines also advocate that efforts should be made to 'mainstream' conflict-sensitive skills throughout development programs, particularly in fragile communities.

It is unlikely that the majority of development assistance programs will be converted to work directly ON the root causes of conflict. It is important, however, to ensure that the engagement of outside actors is conflict-sensitive so that programs are consciously designed to work IN a conflict and not AROUND it.

DAC studies show that the influence of aid in fragile communities can be significant it can exacerbate community tensions and do harm if special care is not taken.

There should be concrete measures aimed at improving the understanding and skills for conflict-sensitive programming. This should involve the conduct Peace and Conflict Impact Assessments, as well as to identify and design conflict-sensitive options and programs on behalf of the Oil & Gas companies.

Analysis, which reflects the inputs and priorities of local actors, is the optimum approach.

The aim is to facilitate the design of conflict-sensitive approaches to potentially fragile communities. Through an assessment of a community profile, and impact profiles, Oil & Gas companies concerned with host community development can make strategic choices and define entry points for engagement which are sensitive to the rich tapestry or relationships and undercurrents which exist in the community.

The Intervention Process

Commitment to intervention on the surface challenges the strongly held view, particularly as it concerns vulnerable rural communities that their peace and security is best preserved by non-interference in communal conflicts. It is our view that institutionalising collective intervention, rather than any unilateral intervention could achieve better results.

The entire question of intervention to prevent a breach or possible breach of peace and security pre-supposes a co-ordinated infrastructure of integrated military, intelligence and institution building capacities.

These are complex matters. If intervention takes place in an atmosphere of lawlessness, fractionalisation and weak governance, what institutions can be deployed before and after a cease-fire to move the conflicting parties from contained tension to manage, confidence building collaboration?

This is a role that the proposed Conflict Management and Security Agency COMSA can play and seeks to play on behalf of multinational corporations, Government and other stakeholders living or operating in the Niger Delta region of Nigeria.

To appreciate the complexity of Niger Delta conflicts, it could be helpful to analyse the various stages of a conflict process:

Stage 1: The stable situation, where there is social stability and adherence to authority and political legitimacy

Stage 2: The build-up of tension, involving systemic social strains, socio-political cleavages and fractionalisation (emerging crisis in the horizon)

Stage 3: Early stages of violent sustainable conflict, involving weakening of authority, structures and governance

Stage 4: Low intensity conflict, involving open hostility, insurgency, counter-insurgency, Government crackdown, humanitarian crises, etc

Stage 5: High intensity conflict involving highly intense anti-social activities, kidnapping, vandalisation, direct and organised combat, massive casualties and killings, population displacement, refugees, destruction of civil society

It is obvious that intervention is not necessary in stage 1. It is stage 2 that is the most delicate. Early intervention is what is required, if intervention is deemed necessary in the face of a clear and present danger.

Early intervention may even be more effective in avoiding the crises of stages 3, 4 and 5 and simply cheaper in human and material costs for all stakeholders.

The really difficult issue is the possible abuse of early interventions by forces considered external to the issues. The solution lies in ensuring the establishment of peace-making institutions in addition to courts, the police and security forces.

The proposed organisation COMSA will have representatives of government, multinational companies, opinion leaders of the communities, and other stakeholders in the region. COMSA will represent the establishment of a specialised organisation for conflict management in the Niger Delta region that will provide early intervention, peace resolution initiatives, as well as provide for the victims of the conflict through the endowment.

It is thus proposed that the private sector, stakeholder governments, Government agencies, concerned individuals, International foundations and donor agencies fund COMSA.

In addition to its conflict management and peacekeeping objectives, *COMSA will be involved in security initiatives that can usher in lasting peace in the Niger Delta region and protect lives and property*. Key stakeholders in the region can be represented on the board of the agency.

Human Rights Issues

There seems to be a direct correlation between Human Rights violations and anti-social behavior (kidnappings, pipeline vandalisation etc). Reducing the Human Rights violations reduces anti-social behavior, as well as contributing to sustainable peace and economic development by creating a better enabling environment in the Niger Delta.

There should be a formal mechanism was in place to assist the oil company in assessing potential Human Rights violations in the host communities where it operates. A cost/benefit analysis will eventually indicate that there is enormous logic to addressing Human rights violation in the region in the short term.

Given the complex realities of the Niger Delta-community dissatisfaction, weak and unresponsive government, security force abuses, and inter-community violence fueled, in part, by oil company and government resources-a new approaches to the problems in the oil producing communities is needed. Respect for human rights has hardly improved there since 1999, despite the presence of a civilian government and the public commitment by many of the oil companies working in Nigeria (especially Shell) to improved engagement with issues of corporate social responsibility.

Local and state governments should be held fully accountable for their inability or unwillingness to effectively utilize revenues, and the federal government should seek to achieve a negotiated solution to the fundamental demands of the peoples who live in the oil producing areas of Nigeria.

Oil companies should broadly assess their interactions with the communities where they work, including employment policies, relations with the government authorities and security forces, community giving, and community relations generally, in order to ensure that they are not exacerbating problems in the delta. Given multiple failures by the bodies involved to fulfill their obligations adequately, a new approach is needed The role of the international community has not been as forceful as it could, or should be.

My proposal is thus calling for the establishment of an independent and specialized agency financed by grants, the NNPC, NDDC and the Oil &Gas companies that will conduct Human Right Impact Assessment" for each new project or facility embarked upon by the Oil & Gas companies.

Such assessments should assess any potential human rights problems related to security arrangements and the potential for creating or exacerbating conflicts that could lead to human rights abuses; and develop plans to mitigate any identified risks. If such an assessment concludes that the human rights risks cannot be adequately mitigated, then the companies should consider whether it is feasible to continue development of those facilities or projects under such circumstances.

he proposed Agency will work with members of the Oil Producers Trade Section (OPTS) of the Nigerian Stock Exchange, to jointly undertake a review of the policy of providing development projects and other benefits only to "host communities," with a view to providing development assistance in a way that reaches a larger part of the population and does not exacerbate local tensions.

This Proposal is aimed at improving the understanding and skills for Human Rights-sensitive programming and contributes to the sustenance of peace and stability in region

NINETEEN

A New Model for Conflict Management in the Niger Delta

This model is aimed at improving the understanding and skills for conflict-sensitive programming especially in the Niger Delta. It will assist in preparing participants to conduct Peace and Conflict Impact Assessments, as well as to identify and design conflict-sensitive options and programs.

It is designed for those who wish to ensure that the impact of their engagement will, as a minimum 'do no harm', and as an optimum, have a positive effect on the conflict dynamics of the community in which the project is taking place.

This model is intended for oil & gas operators in the Niger Delta region, but also applies to other actors (i.e. governments, the political class, security agencies etc,) to identify possible areas for action. The model is ideally used in a workshop setting, but may also be used as a guide for a mission assessment, or working alone. The quality of the analysis depends very much on the individuals or groups that have been assembled, and the questions one asks.

Analysis, which reflects the inputs and priorities of local actors, is the optimum approach.

The aim of this model is to facilitate the design of conflict-sensitive approaches to potentially fragile communities. Through an assessment of a community profile, and impact profiles, development practitioners will be able to make strategic choices and define entry points for engagement which are sensitive to the rich tapestry or relationships and undercurrents which exist in every community.

This model is divided into 3 Parts. Each step has within it an identified objective, definitions of terms used in the tables, questions to stimulate discussion, and an accompanying table.

Together, these steps help complete the Peace and Conflict Impact Assessment Framework. The various steps are briefly described below:

Part 1 consists of the Profile Tools to help users understand the underlying currents and the context in which they work

Part 2 consists of the Impact Tools that allow for the assessment of the possible impacts of engagement in order to help users consider the causes and effects that may lead to unintended negative impacts, and identify unforeseen opportunities

Part 3 provides Decision Tools to consolidate the unintended impacts of a project, and to identify how the project can address the harm or pursue a new opportunity to benefit people.

This model serves as a stand-alone version of material that has been assembled from many sources

PART 1: Community Profile
Objective –
To stimulate discussion amongst those who are planning to engage with potentially fragile communities in order to develop an understanding of their various components and undercurrents.

Rationale –
Profile Tools can be used at any time before a project commences or when it is operational. They aim to strengthen the understanding of the context in which participants work. The Profile Tools use Political, Economic, Social/ Cultural, Security, and Regional/ International lenses.

In order to analyse a community profile, three areas need to be tackled:

(a) what are the issues (indicators) that underpin and drive community tensions?

(b) What are the factors (indicators) that put a brake on rising tensions and serve as the basis for peace?

(c) Who are the main?

Stakeholders involved in the community?

Conflict indicators can be identified at various levels (manifestations, proximate and root causes of conflict).

Similarly, peace indicators can be identified at various levels (ongoing peace efforts, structures and processes in place, and peace-building gaps).

Stakeholder dynamics can be understood by reviewing actions, agendas/ needs, and alliances.

Assumptions

This Profile Tool is a 'light' version of conflict analysis and is not an in-depth conflict diagnostic. It is understood, that assessing the context of underlying dynamics in the community is a prerequisite for determining the impact of projects (i. e. water, agriculture). It is necessary to do at least a light version of a conflict- peace analysis for this purpose.

Political Lens –

Development has traditionally considered itself politically agnostic and has avoided political partisanship; however, extensive studies, including those of the OECD/ DAC Task Force on Conflict, Peace and Development indicate that development at all times, has a political impact, whether intended or not. As a result, actors are moving to more deliberate consideration of political impacts in order to assess all the relevant issues that may affect the success of the project.

The Political Profile should consider the political and social groups in the community, political power and discrimination, and political rights and freedoms.

Economic, Social and Cultural Lens

Economic and social developments are traditional comfort zones for development projects and recently have become more holistic in their approaches. Rather than focusing on one sector, relationships and synergies amongst these sectors are recognised.

Even with new approaches, however, there are unintended impacts across sectors to consider.

External actors need to be prepared to look at impacts that are outside the intended scope of the project. A profile in this lens considers economic assets and deficits in the community, social attitudes, cultural practices, and coping mechanisms.

Security Lens

The security situation in a community can hinder the success of new initiatives if they are not understood. Conversely, the introduction of new resources into communities that are resource- hungry can cause tensions if not handled carefully.

Consideration of security issues can help external actors think about whether their interventions will strengthen or weaken the security of individuals or groups in the community. A profile in this lens should consider inter- community conflict, conflict between groups in the community, and the ability of the community to resolve conflicts.

Rights & Responsibilities

Once a profile of the community has been conducted, it is possible to reassemble the information to identify the key issues that need to be considered, analyse the actions, attitudes and structures that support the concern, and identify who or what is responsible for the situation.

Scenarios & Objectives

These-are developed by assessing trends in key conflict/ peace indicators, as well as amongst stakeholders.

Once these trends are understood, it is possible to make a judgment on where "things are going" by weighing up conflict and peace indicators, and stakeholder developments.

The additional value of scenarios is that they are easily translated into overall objectives, thus "rooting" project objectives in reality. As such, an optimal objective can focus on realizing a best- case scenario and contingency objectives focused on avoiding and being prepared for a worst- case scenario.

STEP.1: Conflict Profile

Objective

To understand the history of tensions in the community, their causes, and what fuels them; to identify the priority issues (root causes) of the tensions and identify the priorities for action.

Definitions

Manifestations: Easily identifiable occurrences (what you see) that indicate unrest in the society. Examples may be civil unrest, high, unemployment, marginalisation of ethnic or religious groups, refugees and internally displaced persons fleeing, corruption, etc.

Proximate Causes: Factors that accentuate and make more severe the underlying causes of conflict. They can support or create the conditions for violent conflict, and are time- wise closer to the outbreak of armed violence. They may change over time. Examples may be poor personal security, availability of weapons, increase in the poverty level, shocks, etc.

Root Causes: Structural or underlying causes of conflict. They are necessary, but not sufficient, causes of violence, and are mostly static, changing slowly over time. Examples may be poor governance, absence of the rule of law, lack of respect for fundamental rights, ethnic diversity, colonial history, etc.

Conflict Synergies: There is no single cause of a conflict. Factors vary in importance and can reinforce each other. Conflict analysis must involve assessing the relative importance of various conflict factors and their interrelationship. The combined effect of conflict factors produces an effect that enhances or reinforces the effect of individual conflict factors.

Process

Identify a manifestation of tensions first, then burrow down through proximate and root causes.

Repeat this process until sufficient information is available to get a broad overview of the context of tensions in the community. Identify synergies last, as reinforcing relationships become apparent.

Key Questions for Manifestations:

- What are the indicators of tension in the community?
- What are the stated reasons for the tensions or conflict?
- Are there tensions within the community or between communities?
- Are there indications of civil unrest, high unemployment, corruption?
- How do these tensions directly impact community members?
- Are there groups that face political, economic or social discrimination?
- Are people leaving their homes because of rising violence?
- Do the indicators selected reflect the concerns of various sectors of the population (women, elderly, poor, children, rich) and the vulnerable?

Key Questions for Proximate Causes

- What are the factors that give rise to, or support the tensions?

- How have existing political processes and institutions fuelled tensions?

- What are the mechanisms that people use to voice their political views?

- How is competition for resources managed? What inequities exist?

- To what extent is identity manipulated for political or economic gain?

- What legal institutions, formal or informal, including dispute resolution mechanisms exist? Have they played a role in the tensions?

- Is the delivery of social services declining or improving?

- Are there systems that support the availability of small arms?

Key Questions for Root Causes

- Legitimacy of the state   Does the community participate politically in fair elections?

    - What is the level of citizen representation or degree of decentralisation?

- Rule of Law How strong is the judicial system?

    - Does the law protect people equally and fairly?

    - Do they have rights to a fair trial that treats them as innocent until proven guilty? Is there biased law application and enforcement?

    - Does civilian powers control the military system?

- Respect for fundamental rights

    - Is there evidence of social exclusion or marginalisation of ethnic groups,

    - Are political, civil and religious rights respected?

- Active civil society and media _ How free are people to express their political or ideological opinions or practice the religion of their choice?

- How free are people to gather to share ideas or form groups?
- Are effective dispute resolution mechanisms absent?

• Sound economic management _ Are inequities related to particular identity groups?

- Are there unique historical legacies, or issues surrounding the distribution of economic, social, or political resource?

STEP.2: Peace Prolife

Objective

To understand what factors can contribute to a sustained peace, reduce the incidence of violence, or prevent the outbreak of violent conflict.

Definitions

Ongoing Peace Efforts: Easily identifiable manifestations or occurrences (what you see, the evidence) that indicate that non- violent solutions are being sought, i. e. groups advocating non- violence, media promoting tolerance, etc.

Peace Structures and Processes in Place: Structures or processes in place for dealing with unrest or violence, and sustaining peace may include: traditional courts, inter- village meetings, a process where elders meet, etc.

The mechanisms put in place specifically for dealing with the conflict may include truth commissions, tribunals, etc., or systemic supports that uphold peace or reduce the "conflict carrying capacity" of society such as the existence of rules governing relations between villages and groups, a culture of tolerance, etc.

Peace building Gaps: Regional or international political, economic, social, and security initiatives requiring attention to sustain peace that are not currently being undertaken either from domestic or external actors. What or who could spoil the peace?

Peace building Synergies: There is no single precondition for sustainable peace. Factors vary in importance and can reinforce each other. Peace analysis must involve assessing the relative importance of the various peace efforts and opportunities and their interrelationships. The combined effect of a number of peace factors can produce an effect that enhances or reinforces the effect of individual peace factors. Paying attention to peace synergies may identify key targets for support in the pursuit of peace.

Process

Left- to- Right logic applies to this Table. First, identify a manifestation of peace, and then identify whether there are processes or structures in place to support sustainable peace, or if gaps exist. Repeat this process until sufficient information is available. Identify synergies last as the reinforcing relationships become apparent.

Key questions for Ongoing Peace Efforts

- Are there groups seeking non- violence or calling for meetings?
- What is the public media saying? Are there independent, private messaging sources?
- Are there groups calling for negotiations, including civil society?

Key questions for Structures and Processes in Place

- Have parties agreed to demobilise their forces or turn in their arms?
- Is there demonstrated commitment on the part of the major conflicting parties to implement a settlement?
- What are the incentives and disincentives to pursue non-violence? Are central actors getting what they want? How much of a threat to peace
- Are those actors who did not get what they wanted?

- What would it take to placate these interests in the short, medium, and long term?

- What degree of consensus exists among political actors and stakeholders? What is the consensus based upon?

- Have trends emerged during the process of discussions or negotiations? Do these trends have any 'predictive' value?

- Are there processes that have been used which appear to have led to some problem resolution? Are these processes worth repeating?

- What are the forms of conflict resolution, and judicial enforcement relied upon by the community, both legal/ judicial or traditional?

- Have you considered indicators at all levels (local, national, international)?

Key Questions for Peace-building Gaps

Are there peace- promoting initiatives that are not being undertaken that need attention?

- Are there sufficient resources devoted to peace promotion (the positives), or are more energy devoted to the conflict (the negatives)?

- What or who can spoil the peace?

STEP.3: Stakeholder Profile
Objective

To understand the potential and actual motivations of various stakeholders and the actions they may take to further their respective interests.

Definitions

Stakeholders: Primary, secondary, and external parties to the conflict. These actors represent the group's and/ or individuals with a stake in maintaining the conflict and/ or building peace.

Actions: Easily identifiable manifestations or occurrences (what you see, the evidence) of efforts made/ activities undertaken by various stakeholders to promote peace or conflict.

Agendas/needs: The vested interests of key stakeholders in maintaining the conflict or working toward peace - opposing or overlapping requirements affected by the conflict or peace. e. g. access to land for pastoralist groups, or medical supplies for guerrillas. Note: wants are different than needs, and some stakeholders have legitimate needs or grievances against authorities.

Stakeholder Synergies: Actors can vary in importance and reinforce each other. Stakeholder analysis should assess the relative importance of the various actors and interrelationships. The combined effect of stakeholders can produce an effect that enhances, or reinforces, the effect of individual actors. Synergies can exist without being consciously pursued. Paying attention to synergies between the actors may identify key targets for support or preventive action.

Process

First, identify a stakeholder, and then plot their actions, their agendas (what they want), needs, and alliances. Repeat this process until sufficient information is available. Identify synergies last as the reinforcing relationships become apparent.

Key questions for Stakeholders

-   Consider all stakeholders (political, economic, social, and security), and race, colour, tribe, caste, language or ethnic group

-   Consider relevant government, military and civil society leaders as well as communities or groups, and latent actors

-   Do the stakeholders selected reflect the concerns of various sectors of the population (women, elderly, poor, children, rich, etc.)?

-   Are there stakeholders who have no voice or are difficult to hear?

- How do they define themselves? What are the core identity features?

- Who are the real leaders of these groups - politicians, soldiers, religious leaders, intellectuals? Are they representative? Do they hold legitimate authority?

- Have the key actors changed over time?

- Consider stakeholders at all levels (local, national, international)?

- Consider the importance of historic, present, and future stakeholders?

Key questions for Actions

- How do the key actors mobilise (i. e. via political parties, armies)?

- Do they hold political power or are they subject to discrimination?

Key questions for Agendas and Needs

- What are the central interests and incentives of different actors?

Peace agendas:

What visions of peace do the stakeholders have? What kind of peace do they want? What are the main elements of their peace? agendas (land reform, national autonomy)?

- What factions or reformist elements exist within identity groups? Are these groups homogeneous or not? Are there spoilers opposed to peace? How great a threat do they pose?

- What are the principle alignments, and do they conform to major social cleavages? Are they diffuse, shifting or stable? What is their base?

- Did central actors get what they wanted? How much of a threat to peace are those actors who did not get what they wanted?

- Are your stakeholders reflective only of the current phase of the conflict?

Consider whether other phases are relevant.

Key questions for Capacities and Vulnerabilities

What capacities do the stakeholders have to support conflict or peace or otherwise affect it?

- Which individuals/ groups have power/ influence?
- What pressures are they subject to from followers, constituents, or opponents?
- What financial, human, and political resources are available to them?

Look for vulnerabilities as well as capacities.

- Do they have formal or informal arrangements of support for continued conflict or peace?
- Are there synergies amongst stakeholders whether intended or unintended?

STEP 4: Responsibilities And Underlying Causes

Objective

To look holistically at the relationship between conflict, peace, and stakeholder dynamics, and the processes and structures that support them; to identify the focal points for future action.

In some circumstances, particularly when dealing directly with the affected, local communities, the Profile questions raise sensitive issues, and participants may be reluctant to express their concerns. Good judgment needs to be used to determine if the question should be raised publicly or not. Building rapport in the group or finding neutral spaces for dialogue and consensus building is a key aspect of moving discussions forward.

Sometimes it is better to start with less threatening issues and builds group confidence over time.

For external actors looking at the impact of their work, one would not expect local communities to "fill in the boxes." Rather, one might engage in a dialogue with local partners and explores areas of concern.

The information gathered could be assembled later in a report for the use of colleagues.

Ideally, project selection should be based on the priorities of local communities and not those of well- meaning outside actors. Local consultation should also take place during the project identification and planning stage.

Definitions

Issue: refers to the issue or concern that was identified from the Profile Tools e. g. the community's right to freedom from X is being denied.

Actions: refers to the actions or failures of action that has led to this problem.

Attitudes: refers to the attitudes or behaviours that caused the action named above. What human rights concern so these behaviours or attitudes reveal?

Supporting Structures and Processes: refers to what systems or structures cause the behaviours or attitudes. What systems cause, reinforce, enable, or perpetuate these attitudes or behaviours?

Responsibility: refers to which body is responsible for causing or addressing the concern.

Process

Left- right logic applies to this Table. First, identify the important, priority issue of concern. Next identify the relevant actions, attitudes, structures that support, or have led to, this concern. Repeat this process until sufficient information is available to get a broad overview of the key issues and how they are supported or "propped up".

Key questions for tying it Together

- Have you looked at root causes as well as symptoms?

- Have you looked at the peace capacities and not just the tension-producing factors?

- Have you looked at the Dividers and Connectors in the community?

- Have you looked at national as well as regional and international factors?

- Have you considered the situation of those who have little power to voice their concerns, as well as those who are easily heard?

STEP 5: Scenarios And Objectives

Objective

To draw out the best, middle and worst- case scenarios in order to prepare and define realistic objectives for engagement.

Definitions

1. **Scenarios**: Scenarios basically answer the question, "What will happen next?" A time frame (e. g. six months) is normally given on scenarios, as are judgments on their likelihood (e. g. most likely, likely, less likely). Scenarios are developed by assessing trends in indicators (i. e. are they getting stronger or weaker, or are they on the rise or decline?) and among stakeholders, and weighing conflict- indicator trends against peace- indicator and stakeholder trends.

At this stage, one may look at what might trigger a change in the current situation (i. e. the death of a key actor), or what might ignite a change in the current situation.

Best- case Scenario describes the optimal (most positive) outcome of the current situation. It may be based on certain assumptions (e.g. rebels decide to negotiate) about stakeholder actions.

Middle- case Scenario describes a "muddling through" outcome of the current situation. It is largely based on an assumption that the status quo (e. g. fighting between parties) continues.

Worst- case Scenario describes the worst- possible outcome of the current situation. It may be based on assumptions (e.g. government launches a large counter- offensive) of stakeholder actions.

2. **Objectives** provide much- needed strategic direction for responses to conflict. They need to reflect a combination of "ground realities" and response capacities, as well as scenarios.

Optimal Objectives are translations of the best- case scenario (e. g. support the negotiated settlement of the conflict). In essence, it is an objective that will direct efforts to realize the best- case scenario.

Status Quo Objectives reflect the middle (muddle through) case. It is debatable whether an objective for the status quo is needed, since the purpose of engaging in a fragile state is generally accepted to be about promoting the best case and preventing the worst case from happening.

Contingency Objectives seek to ensure that practitioners are prepared for a worst- case scenario and prevent these from happening (e. g. develop preparedness plans and discourage a military counter-offensive).

Process

Using the key conflict, peace, and stakeholder factors and their trends, build the scenarios first (approximately 50 words each), describing the state of affairs if the optimal, status quo, or most negative situation evolved.

Next, define an objective for your engagement (50 words max) for each scenario that specifically addresses the key factors you have identified in the Community Profile (conflict root causes; peace opportunities, capacities, and gaps; stakeholder needs, and synergies).

No objective is required for the middle case, unless your objective is to 'muddle through'.

Key questions

- What are trends in key conflict indicators/ synergies, peace indicators, and stakeholder dynamics?
- Is violence on the rise or decline?
- Are peace initiatives getting stronger or weakening?
- Are stakeholders getting stronger or weaker? Which direction are things going?
- What event might trigger or "tip" the balance towards violence or peace?
- What is your judgment about best, middle, and worst- case scenarios when considering the overall (conflict, peace, stakeholder) picture?
- Given your scenario, what objective for engaging in this community is appropriate and realistic?

PART 2: Impact Assessment

Objective

To help users understand the overall impact of their projects and programs by considering the unintended negative impacts, and unforeseen positive opportunities.

Impact Tools are designed for brainstorming, but can also be used by individuals working alone. Looking through various lenses, consider the potential or actual impact of your intervention on people's rights or lives, and identify "who will benefit?" and "who will not benefit?"

Impact tools can prepare you to capitalise on previously unforeseen opportunities as well as mitigate potential unintended impacts. Once a project is operational; they can also be used to evaluate the impact the project is having.

STEP 6: Political Impact

New projects or initiatives in a community may have an impact on political power structures, political rights and processes, political identity and participation, and empowerment even though they may not have been designed to do that. This can have a disruptive impact on relations in the community, or between communities.

Although development workers have traditionally avoided political partisanship, experience from the field, and OECD/ DAC studies have shown that all aid, at all times has a political impact, whether intended or unintended, on the dynamics within the communities in which the project works. Political impacts need to be considered more deliberately and be clearly recognised as an area for consideration.

Process

Identify the key issues outlined in the questions Repeat this process until sufficient information is available. Identify key issues which received a 'yes' or 'partly' answer and look for synergies and reinforcing relationships amongst these key issues and actors.

Key Questions

Consider whether the project will help or hinder:

- Political identity, protection, freedom, and participation
- The levels of participation by women in political processes
- Rights to nationality and recognition before the law
- Rights to a fair trial, innocence until proven guilty and political asylum
- Freedom of thought, conscience, religion, opinion, and expression
- Rights to assembly, association, and political participation in the power structures
- Consolidation of constructive political relationships between state and civil society

- Traditional authority structures
- Transparency and accountability of public decision- making
- The composition or distribution of political resources within/ between state and civil society
- Inter-group tensions
- Formal or informal political structures and processes - either within the formal arena of institutionalized state politics, or within the informal arena of civil society

STEP 7: Economic, Social, Cultural Impact

Objective

To help users understand the economic, social or cultural impact of their projects and programs by considering the unintended negative impacts, and unforeseen positive opportunities.

The introduction of new projects into a community may affect the economic assets or the vulnerability of individuals or groups in that community.

An irrigation project, on the surface, may appear to be worthwhile, but if the economic benefits of that project flow to, or favour one group over another, it can create tensions.

An assessment of these impacts should include rights essential to livelihood security such as economic well being, nutrition, food, water, health, education, the environment, shelter, and culture.

Projects intended for one sector can have crossover impact on other aspects of the community that is unintended. Resource injections can affect economic markets and people's livelihoods.

Process

Identify the key issues outlined in the questions Repeat this process until sufficient information is available. Identify key issues which received a 'yes' or 'partly' answer and look for synergies and reinforcing relationships amongst these key issues and actors.

Key Questions:

Consider whether the project will help or hinder:

Economic the equitable sharing of project benefits Economic infrastructure Access to scarce natural resources

- Economic independence
- Employment or income generation
- Relative economic status of identity groups
- Reliance on an economy related to violence (e. g. small arms)
- Capacities for individuals and communities to define problems, formulate solutions, or resolve problems?
- The status of indigenous or vulnerable groups?
- Social Inclusion of members from the various communities in decision- making. How can you find those who have no voice?
- The ability of individuals & groups to work together for mutual benefit?
- Positive interaction between groups?
- Building bridges between the different communities
- Constructive communications
- Those promoting tolerance or inclusion
- Social services and health care

Cultural

- The attitudes, systems or structures that lead to, or encourage economic rights violations
- Contact, confidence, common interests, or trust between communities?

STEP 8: Security Impact

Objective

To help users understand the impact of their projects and programs on the security of the community and its members by considering the unintended negative impacts, and unforeseen positive opportunities.

A review of the security impact of projects should include effect on tensions between and within communities, and the capacity for individuals or groups for conflict resolution.

The introduction of new resources into a resource- hungry community can create additional tensions. In fragile communities, aid resources can alter security rights dramatically, and affect power structures and relationships.

Communities often have their own internal balances, working relationships and hierarchies. The disproportionate flow of project benefits to one group may shift power balances and make some groups more vulnerable to others.

Sustainable development is not likely to be achieved unless we address the tensions that divide communities.

Projects which make personal or group security more fragile are likely to fall short of their development goals while energy and attention are focused elsewhere

Process

Identify the key issues outlined in the questions Repeat this process until sufficient information is available. Identify key issues which received a 'yes' or 'partly' answer and look for synergies and reinforcing relationships amongst these key issues and actors.

Key Questions

Consider whether the project will help or hinder:

- The relationships between the community and those with whom there are disagreements
- The community's vulnerability to violence from outside, or their capacity to commit violence against outsiders
- The empowerment of those who commit violence or victims to resist violence
- Making potential victims more or less attractive targets
- The individual or group sense of security (physical, food, violence)
- Capacities to pursue non- violent options
- Strengthening or weakening underlying attitudes or systems and structures that cause violence or security rights violations
- Strengthening or weakening local structures for conflict resolution
- Life, liberty, freedom from slavery & torture, displacement, sexual assault, arbitrary arrest and detention;
- The military/ paramilitary/ criminal environment -directly or directly;
- Political, economic, physical, food, security;
- Environmental degradation, resource scarcity, political manipulation, disinformation, mobilisation and politicisation of identity, etc.;
- The development or consolidation of equity and justice, or the means of providing basic needs.

PART 3: Decision
STEP 9: Decision Tool
Objectives

To look holistically at the relationship between the profile of your community, and the impact your project or programme may have on that community. It is also an opportunity to review identified vulnerabilities and capacities in the community.

The Decision Tool aims to help practitioners move from understanding to action. In this step, participants also look at key strategic issues in order to define possible response strategies.

This is the time to reduce or distil a possibly large volume of issues to a manageable number. This distillation process could reflect 1) the urgency of response needed, 2) the identification of priority, root causes of tensions which multiple ripple effects, or 3) a peace- promoting opportunity, which is absent.

There are often constraints or resistance to change - both internal and external, as well as supports. It is important to identify both the obstacles and opportunities in order to decide on an effective course of action.

Definitions

Constraints are attributes that make your decisions more difficult.

Supports are attributes that make the decision easier.

Internal refers to attributes of your own organisation or project.

External refers to those forces outside the organisation that might oppose or support a change. Paying attention to these obstacle and opportunities may identify key targets for action or partners and allies.

Process

Identify the key issues outlined in the questions Identify many strategies for action to deal with these issues (10 words max). Strategies could include something to support, contain, prevent, or be a new initiative in the project. Continue plotting ideas until a broad range of possible responses have been identified. Do not allow your own capacity to respond biases your recommendations for action. If you are a hammer, do not look only for nails. Proposed action could include actors other than your own organisation.

Key questions

- In view of the full analysis, are your identified key issues complete?

- Have you heard the voices of all the stakeholders? Not everyone has the power to speak, but they need to be heard.

- Once the possible strategies for action have been identified, they need to be looked at in terms of: Overall conflict- sensitive objectives - Coherence of the strategy

It is not possible for every actor to tackle every issue. Capacity and resources are usually limited. Choices need to be made. In order to make strategic choices, there is a need to assess the initiatives of other agencies and the capacity of one's own agency in the different fields (governance, economics, socio- cultural and security).

Key questions include:

- What peace- promoting initiatives are being undertaken?

- What is my agency's comparative advantage and capacity?

- Specifically look at your capacity in various fields (political, economic, social, security) at all levels (local, regional and international). What can? be mobilized to impact on the conflict- sensitivity of your project?

- Should you implement policies and practices for more- inclusive participation, or are such efforts adequately supported?

- What are the most critical activities that need to be undertaken?

- What initiatives need to be taken which might enable other things to happen?

- What stakeholders need to be supported to move the agenda forward? What stakeholders need to be contained or included? Are women or vulnerable groups being included or heard?

- Do some actors, either local or external, have a special capacity to respond?

- What is the rationale for the specific initiative being recommended?

- Which agency or group has the greatest capacity to respond?

- What mechanisms need to be put in place to promote collaboration or coordination amongst external actors such as donors?

- Concretely address an unintended harm, caused by the project or identify a new opportunity to benefit the people.

TWENTY

Special Community Investment Trust Funds

The Federal Government should permit Oil & Gas companies operating in oil-producing communities to assign a proportion of equity directly to the communities, to supplement the indirect benefits they receive through Federal taxes and royalties. These funds can go into special trust funds to fund community projects. The source of funds should not be restricted to the Oil companies alone but cover all relevant stakeholders in the area. This has been done successfully in other parts of Africa; for example the Bafokeng tribe who share in the platinum resources from the platinum mines in South Africa.

This grass-roots funding mechanism at the level of the host communities through the establishment of Special Investment Trusts can incorporated by the communities themselves (with the aid of consultants and NGOs acting in an advisory capacity) in which 4 different partnership groups (P-1 to P-4) will invest funds on behalf of the communities and will have representation on the board of trustees:

P-1: Government agencies i.e. the NDDC and other State development agencies

P-2: Donor Agencies (including the ADB)

P-3: The Oil & Gas Companies

P-4: The Business Community representing the private sector.

The implementation arrangements for the Trust are outlined in the diagram below. Under the terms of the Trust deed, the Trustees are responsible for oversight of the management of the Trust's assets in accordance with the deed. The deed defines the purpose of the Trust and the way in which the trustees' responsibilities are to be discharged. Aside from the general principles set out in the trust deed, the trustees will be obligated to ensure that the trust's investment guidelines and policies and procedures are followed.

The trustees will be mandated to seek technical guidance from a programme investment committee (PIC) and to implement the programme using the services of a technical manager and finance manager. As a result the role of the trustees is primarily one of oversight, ensuring that the systems and processes necessary to ensure programmatic and financial integrity are followed.

The finance manager will be required to undertake the day-to-day financial management and administration of the trust, reporting directly to the Trustees. In order to minimise the involvement of the trustees in ensuring that the terms of the Trust are met, the finance manager functions will be formally separated from those of the technical manager responsible for programming of the Trust's development activities.

The finance manager will work alongside the technical manager but be legally directly accountable to the trustees. A full-time presence is required in the Trust office to ensure close day-to-day co-ordination with the technical Director in implementing the Trust's programme.

Trust Structure

Scope of the Work of the Trust Fund Technical Manager:

- Development and presentation to the Trustees for approval of a three-year strategy plan.
- Development and presentation to the PMC for approval of an annual plan outlining inputs, activities and outputs required to achieve the business plan;

- Integration of on-going partner's support activities, where consistent with the trust's objectives into new programmes;

Investment cycle:

- Identification of new investment opportunities in support of the strategy to develop the Community;

- Appraise/undertake due diligence of promising investment opportunities;

- Present investment proposals to quarterly PIC meetings for opinion to the trustees (nb: all documents for the PIC must be submitted at least 14 working days before scheduled quarterly meetings, later submissions will be at the discretion of the PIC);

- Ensure the trustees receive full documentation of investment proposals, approved minutes of PIC deliberations and PIC opinions on proposals (including objection, qualified objection and no objection);

- Presentation of documentation to the finance manager to facilitate preparation of funding Agreements and disbursements to implement projects approved by the trustees;

- On-going strategic inputs and technical support to investments in order to maximize developmental returns (including participation in boards where appropriate);

- Monitoring of investments, based on international best practice, to ensure effective tracking of institutional performance;

- Ensure rapid action to suspend disbursements/payments (in conjunction with the Finance Manager) in case of under-performance or other non-compliance against terms of funding agreements or contracts;

- Manage appropriate exits from investment relationships (including identification and co-ordination with potential new institutional investors).

- Provide strategic inputs and engage with policy level work on Community development involving Government, development partners and others (notably taking responsibility for feeding experience from field level work into higher level policy debates);

- Liaise closely with international technical experts

- Enhance opportunities for Community development based on service provider capacity by assessing options for local procurement of all proposed, direct or indirect, sub-contracts and related opportunities;

- Co-ordinate with technical experts in promoting local service procurement with other donors and investors;

- Identify and exploit opportunities for joint action with other donor programmes and investors to maximize impact on Community ;

- Maintain global links with sources of emerging best practice

Operations and administration

- Support the efficient and effective operation of the trust in a manner at all time consistent with the terms of the trust deed, funding agreements with donors and the defined policies and procedures;

- Develop appropriate investment mechanisms/contracts, in conjunction with the finance manager and trustees, to deliver support to Community development projects;

- Work with the finance manager to ensure appropriate reporting mechanisms and accounting systems to allow efficient accounting to the trustees and funders and other investors in the Trust Fund;

- Support the finance manager in ensuring project partners, consultants and all other suppliers comply with all specified reporting requirements, funding conditions and contractual terms under funding agreements and contracts;

- Specifically providing explicit authorization to the finance manager for all payments including any disbursement requests that the Finance Manager intends to submit to the Trustees;

- Working with the finance manager ensure contracting of service providers to support Trust Fund activities is undertaken in a competitive manner, and consistent with the procurement policy of the trust;

- Prepare and submit management reports to the PMC in a timely fashion.

- Collaborate closely with the finance manager in the timely preparation of quarterly financial reports, providing clear analysis of variances against budget;

- Support the finance manager in preparation of three year, annual and quarterly financial budgets and cash-flow forecasts;

- Assess and approve the annual budget and cash-flow forecast produced by the financial manager;

- Co-ordinate closely with the finance manager in ensuring timely funding requests are made to donor partners to ensure donor liquidity

- Development of formal operating Policies and Procedures (in collaboration with the Finance Manager) to be submitted to the PIC for opinion and subsequently to the Trustees for formal adoption.

- In association with the Finance Manager, make proposals for changes to the Policies and Procedures as required, to be submitted to the PIC for opinion and subsequently to the Trustees for formal adoption and update to the Policies and Procedures as appropriate

- Collaborate with the finance manager in the establishment and smooth management of a project office;

- Identify all potential, actual or perceived conflicts of interest and present to PIC for opinion to the trustees;

Support to Trustees and Finance Manager

- Provide all reasonable support to the Trustees in executing their obligations under the Trust deed, providing information expeditiously where requested and advice on specific issues raised.

- Offer close operational support to the Finance Manager in support of effective and efficient execution of the Trust's Community development programme. The Technical Manager and Finance Manager are expected to work as an integrated team with the objective of supporting the Trustees in achieving the purposes of the Trust.

- The Technical Manager will be obligated to bring to the attention of the Trustees and PIC all potential, actual or perceived threats of financial loss, violation of defined operating policies and procedures or other material threats to the Trust.

Conduct of the work:

The technical manager will be legally accountable to the trustees, who are mandated under the Trust deed to evaluate the technical performance of the manager seeking the opinion of the Programme Investment Committee (PIC). The technical manager formally reports to the Trustees. The Programme Investment Committee (PIC) will provide advice to the Trustees on the technical performance of the technical manager and will meet on a regular basis, at least quarterly.

While the specific approach to meeting the scope of work defined here is left to the consultants to determine (notably in relation to operational and administrative support), it is necessary that a full-time, qualified, resident manager be provided to act as principle point of contact for all work and to provide the lead on technical inputs. Proposals for approaches to fulfilling other aspects of the scope of work should be outlined concisely in the response to these terms of reference. Options might include sub-contracting of functions, engagement of additional support staff or drawing on existing corporate resources.

Examples of Trusts in Other Regions

There are examples of trust funds in other parts of the world.

These include:

- Heritage Savings Fund Alberta, Canada

- Copper Fund Chile

- Reserve Fund for Future Generations, Kuwait

- State Petroleum Fund Norway General Reserve Fund Oman Alaska Permanent Fund Alaska, USA Stabilization Investment, Fund, Venezuela Future Generations Fund Chad

- Alaska has an oil fund made up of accruals from oil rents and royalties, which cushions the state's economy against cyclical shocks in earnings. The Nunavut Trust in Canada gives an example of how revenues can be shared with local. communities, and also how such communities can be compensated for negative external consequences of oil and mineral extraction.

The Alaska Trust

The Alaska Permanent Fund is one of the oldest natural resource funds. Established in 1976, it has grown to US $25 billion. Citizens have been extensively engaged in public consultation; each year they receive dividends.

The Alaska Permanent Fund Corporation manages the fund, which is invested in stocks, bonds and real estate. Unreserved assets (income from the reserved assets) are retained in the Fund until appropriated by the legislature. As of 30 June 2003, the unreserved assets totaled US $100 million. To date, the legislature has used these assets for three primary purposes:

- Payment of dividends (US $12.5 billion from 1982 through 2003 paid out to Alaskans)

- Inflation-proofing the reserved assets

- Increasing the size of the reserved assets

Critics have argued that Alaska's fiscal policy should include the use of a portion of unreserved assets to support state services and programs. Paying out the same dividend to everyone does not take account of varying social or economic needs. On the other hand, the system ensures that the Government does not misuse the oil wealth. Transparency underpins the fund's operations. The corporation publishes frequent reports and its Web site contains comprehensive information about legislation and governance.

Nunavut Trust in Canada

The Nunavut Trust illustrates not only how revenues can be shared, but also how a community can be compensated for the negative external consequences of oil and mineral extraction, and can turn this compensation into local development.

Nunavut is the northernmost territory in Canada, bordered on the east by Greenland. The Trust was set up in 1993 as part of a land claim settlement between the Canadian Government and the native people of Nunavut.

The settlement recognized native rights to land and resources, and reflected the right of native people to participate in decision-making concerning the use of such resources, including offshore.

The settlement included a share of royalty payments from natural resource development in Nunavut. The Trust administers this as well as a compensation payment from the Canadian government for the negative local consequences of resource extraction. Some US $1.2 billion dollars in compensation money will pass from the federal government to the people of Nunavut over 14 years, ending in 2007.

The Nunavut Trust is unique because its beneficiaries run it independently of any Government. It is a community-managed fund. Three regional Inuit civil society norganisations—Kitikmeot Inuit Association, Kivalliq (Keewatin) Inuit Association and Baffin Regional Inuit Association—appoint the trustees.

Unlike other funds, the Trust does not pay out dividends. Its designers feared that such a payment could lead to increased unhealthy spending on personal consumption such as alcohol. Separate non-profit organizations are responsible for spending the money, the largest of them being Nunavut Tunngavik Inc. Money has gone into health, sports, local business development, etc. Nunavut Tunngavik Inc. meets regularly with Inuit communities to discuss priorities for spending, so the locals appear to be heavily involved in the governance of the Trust.

TWENTY-ONE

Building Strategic Partnerships & Alliances

The expansion of basic services and the development of sustainable infrastructure are key challenges in the Niger Delta. One option of expanding service delivery is to enhance the role of the private sector in the financing and management of basic services such as water, sanitation, waste management, road maintenance, etc. Consequently, Public Private Partnerships (PPPs) are increasingly viewed as a mechanism to provide state and municipal (local authorities) services on a cost effective and sustainable basis.

Faced with the above reality, the author has realized the following:

- There is a need for higher levels of private, local and foreign investment in expanding public infrastructure in the Niger Delta;

- There are opportunities to leverage more private finance through "Public Private Partnership" financing mechanisms;

- Collaboration between the public and private sectors is beneficial in the soliciting of support for sound projects;

- Staff in public and private institutions need to be trained to assess investment proposals received from private financiers;

- Strong, supportive, and stable regulatory governance structures increase the chances for success of PPP's;

- The success of PPP's often depends of concerted regional efforts and effective cross-border initiatives.

176

- In order to give impetus to the implementation of PPP projects Government must recognize that an overall conducive environment necessary for the facilitation of PPP projects needs to be created. It is within this background that we are proposing that the Oil companies and if necessary Donor Agencies and other Private Sector organizations finance a full review of the prospects for PPPs in the Niger Delta and catalyse the implementation of its findings

A special committee can be set up to review ways to stimulate greater efforts to promote the implementation of PPP projects through various policy documents of the state governments in the Niger Delta.

Development Context

As earlier stated, investment is key to economic growth that is necessary for sustainable poverty reduction. Developing countries often do not have adequate local savings to support a level of investment sufficient to generate enough economic growth to reduce poverty; therefore, they need to be able to attract foreign investment.

The professionals and development experts have recognized the need for higher levels of foreign investment, especially in expanding public infrastructure, to support sustainable poverty reduction.

Mobilizing domestic private investment into longer-term infrastructure projects is a further objective. The provision of physical infrastructure itself can have a dramatically positive impact on poverty by lowering family expenditures on basic services, especially on water and energy, and reducing the barriers that prevent the poor being able to respond to market reforms (e.g. farmers with road access to markets are more likely to take advantage of price changes for their products).

Besides increasing the overall volume of investment, foreign investment also frees up scarce public resources for use in areas with higher social returns, such as for health and education services, and, if directed towards expanding public goods, for instance by providing transportation, electricity or communications infrastructure, can create opportunities for further investment by indigenous entrepreneurs in other sectors of the economy.

The envisaged impact of the project over a five-year period will be:

- Increased support for Niger Delta regional efforts to promote the level of domestic and foreign direct investment that in turn would increase the number of projects financed through PPP's.

- Processes to revise policies will be underway in several States in the region to facilitate an environment conducive for PPP's.

- Trained men and women practitioners in public and private sectors will actively seek to introduce PPP investment projects in the Niger Delta region.

- Sustainable PPP capacity building opportunities available to men and women from the Niger Delta region.

Institutional Environment In The Niger Delta.

The success of the concept of attracting investment through Public Private Partnerships (PPPs) depends on the extent to which the strategic framework required for the implementation of PPP's is in place.

It has been observed that while there is no overall guiding policy on PPP's in Nigeria, the potential for such projects is quite apparent. Initiatives in this regard are with government departments which are independently planning to start their PPP pilot projects, ranging from the provision of office accommodation, primary hospitals, IT facilities and toll roads to fleet management services. There is no doubt that the approach to PPP implementation may be varied and possibly fragmented, unless a uniform and guided approach is subscribed to.

There is therefore need for the formulation of an overall policy environment for PPP's as a matter of urgency. A review of the legislative, institutional and regulatory environment also needs to be considered as a first step to facilitating the successful implementation of PPPs, so that the strategic framework required is developed on the basis of the findings and recommendations from the review.

The proposed review should include an analysis and an assessment of the institutional capacity as well as the enabling policy, legal and regulatory environment in the Niger Delta to support PPPs including appropriate reviews of Institutional, legislative and regulatory issues.

The content of the review can cover:

- Official policy view regarding PPPs;
- An overview of key legal and legislative instruments (to what extent do they provide a conducive environment for PPPs or serve as a barrier?)
- Key changes/developments in the past few years;
- Institutional review relevant to PPPs (central coordinating institutions, line function responsibilities);
- Establish potential conflict of current regulation with PPP projects;
- The impact and quality of current regulation;
- Review current public financial management processes and procedures and establish possible impact on PPPs;
- Implications of current environment for public policy and future projects;
- Perceptions of institutional/legal environment by various stakeholders;
- Comparison of the PPP methodology with traditional public procurement mechanisms;
- Detailed recommendations for changes to the legislative, regulatory and institutional frameworks to make the environment conducive to PPP implementation.
- Establish general capacity in public sector to implement and manage PPPs;
- Recommendations to the legislative, regulatory and institutional frameworks to make the environment conducive to PPP implementation;

- Relevance of findings to key projects already implemented/ under consideration;
- Propose strategic implementation framework for PPP implementation in the Niger Delta region.
- Propose general guidelines for PPP procurement, monitoring and evaluation in line with the assessment

Tri-sector Partnerships

Given the complex realities of the Niger Delta-community dissatisfaction, and inter-community violence fueled, in part, by oil company and government resources-a more comprehensive approach is needed. One can explore the role of tri-sector partnerships in providing answers to the unresolved social management and sustainable development challenges confronting non-renewable natural resource (oil & gas) projects. This initiative looks for creative ways in which these partnerships can:

- Promote more equitable and visible economic development in the region of operation;
- 'Pool' resources, skills and experience to increase the quality, reach and sustainability of local public services - health, education, water supply, housing water disposal etc;
- Encourage local business activity and leave an economic legacy independent of the oil, gas or mining business;
- Improve the quality of resettlement and income restoration programmes;
- Deliver effective community development in situations of violent conflict;
- Overcome weak capacities in civil society and local government.

To this end the programme can work with project and local stakeholders to:

- Identify where tri-sector partnerships might contribute the greatest **'added value'** to what the parties can achieve alone;

- Facilitate agreement between prospective partners on objectives, roles and responsibilities in areas such as health care, education, employment, business enterprise, training and environmental management;
- Build capacity to enable tri-sector partnerships to be effective and durable;
- And learn lessons from the impact of the partnerships on social investment and sustainable development.

Action Plan for Strategic Partnership in the Niger Delta

Who are the stakeholders for inclusion in the partnership Action Plan?

- Governments (local, state and federal)
- The global community (NGOs, governments, multilateral and bilateral organizations, etc.)
- NGOs and community-based organizations
- The organized private sector
- The citizens of the Niger Delta and other parts of Nigeria
- Oil-producing communities and the rest of Nigeria

Partnerships should be *mutually beneficial*

One should assure that those who have strategic advantages and benefit from the status quo do not lose out when making the concessions required for partnerships. But building mutually beneficial partnerships will determine the future course of development. This can be done through the following actions:

- Governments and communities, and governments and citizens—to nurture peace, good governance
- Communities and oil companies—to encourage peace and sustainable human development
- Inter-governmental relationships (local, state and federal)—to resolve resource control problems, the peace and governance agendas, etc.;

- Intra-governmental (the executive branch, legislature and judiciary)—to build institutions

- International actors and communities, international actors and governments, international actors and NGOs/community-based groups—to build peace, good government and sustainable human development; and

- Inter-community—to cultivate peace;

- NGOs and community-based organizations—to bolster capacity development and public monitoring.

- Coordinate and harmonize development programmes to avoid duplication, conflict and waste.

- Plan for sustainable development, and with the aid of competent facilitators, advise on appropriate policies for symmetrical community development and for training in community development.

- Commit to the judicious use of counterpart funding, with utmost accountability and transparency.

- Create and sustain the enabling environment for meaningful development that can sustain the optimal use of the rich endowment of resources in the Niger Delta region.

- Pursue investment flow promotion, and the participation of stakeholders in making decisions about development planning, programming and budgeting.

- Ensure participation across the various tiers of Government and their agencies in executing the action plan.

- Utilise dialogue and consultation for conflict resolution (e.g. alternative dispute resolution or mediation approaches), towards the achievement of lasting peace and security.

- Empower women and youth groups so they can help themselves and contribute towards development and the public good.

- The NDDC, should catalyse sustainable partnerships in various ways, including through periodic benchmarking of development needs and goals in the Niger Delta, the periodic audit of progress rates in the pursuit of development.

Both the Public and Private sector must be committed to the judicious use of resources, participatory planning, sustainable development and accountability to the people.

- The Private sector must be committed to corporate social responsibilities that add tangible value to the operating environment and help fulfill critical development gaps.

- Promote good social relationships and mutual coexistence with the host communities.

- Shun unscrupulous business practices and the abuse of the environment.

- Encourage consultation and popular participation in supporting community development.

- Ensure synergy in development interventions.

- Exhibit good faith and uphold international best practices in compensation payments and their management.

- Embrace and uphold legitimate avenues for seeking redress for periodic injustices and discontent arising from strained government and/or host community relationships, through consultation, negotiation, judicial remedies, etc.

- Host communities should recognize the need for, and show concern about, the security of petroleum operations through good relations as hosts to the multinational oil and gas industries and other firms.

- Assist in building institutions that support development of host communities, offer mentoring for the acquisition of productive skills and grant appropriate scholarships to aid the development of critical industrial competencies.

- Offer capacity assistance to other stakeholders to better optimize their contributions to development.

- Build the conditions for sustainable peace and an enduring, reasonable socio-political climate— through the creation of social avenues for cordial mutual interfaces, dialogue, open communication and cooperation, among communities and across stakeholders.

The outcomes of partnerships often depend on the balance of strategic advantages: Unbalanced relations tend to produce unbalanced outcomes.

TWENTY-TWO

Mobilising Resources

Government alone cannot effectively reduce poverty and promote sustainable development in the Niger Delta. The private sector should be encouraged to play a greater role, as well as private foundations. Of course international donor agencies are already playing a significant role and additional efforts can be integrated in this current proposal. This proposal focuses on a minimal role for government and a maximal role for the private sector, the donor community and corporate grant-makers/private foundations pooling resources into a new independent funding organisation that for example can be named the *Niger Delta Infrastructure Fund (NDIF)*

Rationale For The Committee And NDIF

This can be a new initiative for funding sustainable development and poverty eradication programs in the Niger Delta by adopting a comprehensive approach in providing for infrastructure development in the Niger Delta region.

In order to avoid duplication of efforts, we are proposing the establishment of a consortium (the committee) comprising representatives from Government and other public sector agencies, the private sector, donor agencies, private foundations, etc to pool resources together under the auspices of the proposed committee, which will act as an umbrella organization to fashion out a common, co-ordinated approach and implement various action plans and programs.

Structure

It is proposed that this new umbrella committee will be co-ordinated by a new independent body, the *Niger Delta Infrastructure Fund (NDIF)* with the assistance of consultants

With head-office in Port Harcourt, it is proposed that each participating organisation will have at least one (1) representative on the main Committee, with the Committee chairman nominated by the consensus or by the vote; the vice-chairman can be nominated by the Private Sector. The main Committee will also be responsible for policy direction as well as determination of the uses of resources at the disposal of the NDIF.

The NDIF will in turn have a Board of Trustees comprising senior officials of participating organizations and distinguished Niger Delta citizens and an Executive Committee.

Invitation for participation can therefore be classified into four (4) groups (with proposed percentage resource contributions):

GROUP A: Public Sector (15%)

* The Federal Ministry of Finance
* The Central Bank of Nigeria
* The Bank of Industry (BOI)
* The Niger Delta Development Commission (NDDC)
* State Governments in the Niger Delta
* The Nigerian National Petroleum Corporation (NNPC)
* The Nigerian Agricultural Cooperative and Rural Development Bank
* Etc.

GROUP B: Private Sector Organisations (50%)

* The Joint-Venture Oil & Gas Companies
* The Banks & Insurance Companies
* The Manufacturing Industry

* Other private sector organisations

GROUP C: International Donor/ Development Agencies (20%)

* The World Bank
* The UNDP
* The EEC
* The AFDB
* DFID
* The CDC
* DEG
* CIDA
* USAID-Nigeria, Etc

GROUP D: Corporate Grant Makers and Foundations (15%)

With the decline in Official Development Assistance (ODA), there is expectation of heightened involvement of private sector investment. As private investment flows now dwarf ODA in many regions of the world economic trends also reveal proliferation of new foundations and corporate grant makers investing in development programs and initiatives.

Private investment, corporate grant making and foundations are becoming a larger part of the development investment picture in Africa. Often these organizations bring resources beyond funding, by way of expertise, commitment, flexibility and creativity.

By building an alliance of the public sector, donor agencies, the private sector, corporate grant makers/foundations, advantage can be taken of creative synergies and leveraging investment. The creation of such an alliance will enable the achievement of outcomes that would not be possible acting alone.

Target Zones:

The Niger Delta region is made up of 9 States. The broad approach should involve zoning the region into 3 zones (3 States per zone), with sub-committees established for each of the 3 zones. However resources constraints might compel the new initiative to implement programmes zone by zone.

Over the past decade, the proportion of people in extreme poverty declined more in some areas of the Niger Delta than in others.

There can be 5 operating sub-committees manned by representatives of the participating organizations:

* Strategy and planning.
* Technical/programmes.
* Financial management.
* Implementation.
* Monitoring & evaluation.

Syndicated Pooling Of Resources:

Resources, particularly financial can be pooled into a common fund with each organization giving instructions on which of the various programs it is interested in funding, during joint meetings of the committee at the proposed committee secretariat in Port Harcourt. Their contributions will be separate and without prejudice to existing commitments they might have or plan to have in the region.

Highly skilled financial professionals will manage the funds raised while awaiting apportionment for specific intervention programs. Funds will only be released on a case-by-case basis for specific intervention programs and in accordance to with the policy objectives.

The Emerging Africa Infrastructure Fund:

The NDIF can pull in funds from the Emerging Africa Infrastructure Fund.

This is a unique public-private financing partnership initiated by the Private Infrastructure Development Group one of whose founding members is the U.K Government's Department for International Development (DFID). The fund represents a new financing approach for the long-term alleviation of poverty in Sub-Saharan Africa through combining public and private funding partners and adopting commercial and developmental principles in support of sustainable development and economic growth.

Development of infrastructure and the involvement of the private sector are the two key points of focus for the fund.

Infrastructure development is seen as a pre-requisite for economic growth and the private sector is seen to be best able to identify and manage risks associated with such development in many sectors so as to ensure assets and services perform over time.

Furthermore, infrastructure in sub-Saharan Africa has suffered acute under-investment with the public sector being increasingly incapacitated and constrained. The Fund therefore comes at a time of profound need in the region.

Inadequacies Of Past Efforts.

Most programs targeted at development in Niger Delta, mainly in rural areas, are essentially empowerment programs and at the same time credit schemes aimed at enhancing their economic lively-hood and halting rural-urban drift. They are thus conceptualised to moving resources to rural areas for productive ventures that benefit the rural poor and low-income groups at large.

However it is pertinent to indicate certain inconsistencies and problems common to all of them. These include, poor executive capacity and management, inadequate funding, high rate of corruption, bureaucratic bottlenecks (top-down), rigid implementation, misuse of credit facilities and low performance of the economic units making repayment almost impossible.

TWENTY-THREE

International Agencies & The Niger Delta

The European Union

The European Union (E.U.) has a €84 million (roughly the same in U.S. dollars) plan for the period 2001 to 2007 covering 5,000 "micro-projects" in the Niger Delta (Bayelsa, Delta, and Rivers States). The projects focus on water supply, the health system, education, rural transport, and income generating schemes, as well as micro finance.

The Cotonou Agreement governing relations between the African, Caribbean and Pacific (ACP) countries and the E.U. includes provisions relating to human rights, democracy, good governance, and the rule of law. Article 96 of the treaty provides that if there is a dispute over human rights issues the parties may request "consultations" about the issue, though this provision has rarely been invoked.

There have been discussions within the E.U. institutions about issues of corporate social responsibility, but no binding or even voluntary standards have been adopted for European corporations at E.U. level.

The World Bank

In addition to the support for the NDDC mentioned above, the World Bank has also made loans to the Nigerian government that may benefit the Niger Delta.

In March 2002, the World Bank approved two loans to the Nigerian government, totaling U.S.$237 million, for the development of the health system and for a community-based urban development project.

190

Akwa Ibom was one of the states selected to benefit from the urban development project.

In June 2001, the International Finance Corporation (IFC), the World Bank's private sector lending arm, approved the establishment of a U.S. $30 million revolving credit facility by the IFC, SPDC and a local bank.

The facility would provide access to credit for Nigerian oil service companies to assist them compete with international contractors. While the IFC did undertake some consultation, this was cursory, and the project is poorly designed in other respects: for example, there are no explicit provisions to ensure that the entrepreneurs benefiting come from the delta area itself nor that the IFC will monitor compliance with environmental and other standards.

In March 2002, the IFC announced a partnership with Nigeria's Fate Foundation, a nonprofit organization, in a project to develop entrepreneurship and small business development in the Niger Delta area.

The World Bank has established a website on "Best Practices in Dealing with the Social Impacts of Hydrocarbon Operations," in collaboration with a group of oil companies and nongovernmental organizations committed to the protection of the environment and mitigation or elimination of any adverse social impacts from oil and gas operations. The website remains a work in progress, however, and there are no mechanisms for monitoring company compliance with the recommendations.

The Bank also conducted a review of its role in the extractive industries, which gives it the potential to take on stronger governance, transparency, and human rights roles. But this review will not be completed until 2003 and it is not clear whether the Bank will take on its recommendations; even if it does, they will take several years to implement.

International Monetary Fund (IMF)

Proper management of the oil revenues at all levels of government-federal, state, and local-is key to solving the problems that beset the

Niger Delta. Since civilian government was restored, the IMF has supplied Nigeria with policy advice, technical assistance, and training, as well as financial support "for policies that will help achieve the country's economic and social objectives.

The G8

In June 2002, the G8 industrialized countries adopted an "Africa Action Plan," by which they agreed to support African leaders' own efforts to overcome obstacles to development in Africa through the New Partnership for Africa's Development (NEPAD). Among the commitments made are many relevant to improving the situation in the Niger Delta, including promoting participatory decision making, supporting the reform of the security sector, combating corruption, and helping attract investment.

Development Funding Utilising The Private Sector

How can development agencies and government in the Niger Delta utilise the private sector as a means of accelerating development and prosperity for its people.

There are two fundamental points. First, the private sector may often be better at assuming performance and market risks than public agencies. Shifting those responsibilities and risks to the private sector may thus be desirable. As part of this shift in risk allocation, the exposure of domestic taxpayers especially among the poor to the burden of public debt can be reduced.

The second, when shifting risks, it is often desirable to un-bundle some of the products of the World Bank Group. For example, by un-bundling the policy risk function from the traditional Bank loan, policy risks guarantees have been created in the Bank and MIGA, that shift commercial risks to private parties.

There may now be further options available to shift performance risks to private parties so as to improve the investment climate and service to the poor. Unbundling the subsidy component embedded in some

World Bank Group financial products would allow targeting it better at its purposes or beneficiaries.

Traditional aid in some sectors has at times been associated with disappointing performance by state-owned agencies that were funded. In addition, subsidies embedded in aid funds may not have reached the poor but benefited the better off. There are thus two main issues: How to improve service to the poor and how to ensure that any subsidy judged necessary actually benefit low-income groups in the Niger Delta.

The key to better service delivery is to shift the performance risk more effectively to service providers (the private sector) and away from taxpayers. This can be achieved wherever it is possible to disburse aid when service is delivered and not when inputs are constructed. Such schemes are in principle, feasible in most parts of infrastructure and some areas of health and education. Service standards should be responsive to beneficiaries by empowering them to choose. This includes community participation to set goals for collective goods and services.

Providers (for profit and/or not-for-profit) can compete for the right to provide service on the best possible terms. Funding can be obtained in the market when the provider is competent and cash flow is expected to be adequate, which is facilitated when aid is disbursed upon achievement of contractual service obligations. The IFC could help to develop the market for this.

Policy risk guarantees by MIGA or the Bank can help deal with the special risks in difficult environments. In many complicated policy environments, small-scale solutions that are within the managerial and financial reach of domestic providers may provide appropriate solutions. (Note that output-based funding could be achieved with grants or with loans. In the case of loans, ultimately, the local taxpayer end up paying, while with grants, the foreign taxpayer pays-both in the case of concessional loans such as IDA).

Normal competitive markets are out-put based schemes. Consumers pay when the product or service is delivered, not when the factory is built.

Typically improved service delivery systems by themselves will help the poor most, because access to service rather than subsidies are crucial for them.

When it is judged that some form of subsidy is needed to help the poor this can be combined with market-type mechanisms. Output-based aid allows in particular one to shift performance risk to private parties, while retaining the option of subsidizing user fees partially or completely. Output-based subsidies can also help reach sub-sovereign projects efficiently without complications of counter-guarantees.

It may be argued that such output-based schemes and targeted subsidy schemes are hard to implement in regions with weak governance systems. They are indeed harder to implement there than in other places. However, it is not obvious why, therefore, potentially less effective alternative approaches should be favoured. The question is: if feasible output-based schemes with competent sponsors, policy risk guarantees and adequate expected cash flow (possibly supported by aid) do not work, then what chances are there that traditional approaches will work.

T W E N T Y - F O U R

Duty Free Tourist Zones

Tourism has emerged as one of the fastest growing economic sectors both internally and in Africa. The Niger Delta region has no reason to be different considering its vast wetlands its cultural richness. Features include adventure tourism, eco-tourism and cultural tourism. Vast under-developed areas exist in the Niger Delta (including heritage sites) with the potential for tourism development.

What is actually needed right now is increased capacity for tourism in terms of 5-Star accommodation outside the urban areas and transport (inland and airlines). There is also tension between managing tourism growth and conservation of the natural resources base, exploitation of which provides the financial resources for infrastructure development.

As a new initiative special Duty-free Tourist Zones can be established in each of the Niger Delta States that have ports. Special areas can be carved out in the port areas for international and local tourist installed with facilities covering:

- New world-class hotels,
- Gambling Casinos,
- A private Golf Club,
- Restaurants
- A Mega Shopping Arcade
- Western style Cinema Halls,
- Business Centres,
- Banks

- Heliport/Helicopter Services
- Special Tourist Bus transport
- Etc

A development strategy can be devised for attracting private investment for the project, thereby enhancing the Niger Delta Tourism industry that would maximise the benefits accruing to Niger Delta State economy.

Potential Benefits

This project, properly designed can be of major significance to the development of the Niger Delta that is facing a major rehabilitation and reconstruction programme. The importance of lies in its ability to:

- Provide Employment: The project will provide job opportunities either directly or indirectly and can represent a significant proportion of the Delta State workforce.

- Rural Development: Most of the Niger Delta's primary tourist areas are largely located in areas lacking infrastructure development. Consequently, tourism represents the only for of employment for a large section of the community. Moreover, tourism in these areas has an important effect on the cash economy of local peasant farmers.

- Generate Scarce Foreign Exchange: Shortage of foreign exchange represents a major constraint to development. Properly managed, tourism can make a significant contribution to regions foreign exchange income.

- Government Revenue: Tourism can be a source of receipt for the State Governments Treasury. Because of the narrow tax base in many States of the Niger Delta, tourism through various levies can significantly boost internally generated revenue.

In such a complex region like the Niger Delta State with contrast of cultures and beauties of Nature, tourism can play an important role to advance reciprocal understanding and mutual respect and thus act as a vehicle for enhancing unity in the region.

TWENTY-FIVE

Value-Added Social Investments

Greater demands are being placed on companies by governments, NGOs and the general public with regard to their social obligations to their host communities.

My advice to these companies is:

Determine your "magnetic north"

Debate and dialogue

Determine your legacy

Be a good guest

Put employees first

Know your neighbour

Handle with care

Pursue smart partnerships

Build-in the strategic business case

Focus on individuals

Reputation matters

Defining the issue: an important first step.

"Corporate social responsibility is the commitment of business to contribute to sustainable economic development, working with employees, their families, the local community and society at large to improve their quality of life."

The common saying that health is wealth can be given practical expression by the following example. A practical example of how companies operating in the Niger Delta can exhibit their corporate social responsibility is by sponsoring a Nutrition Initiatives.

The aim of this nutrition program is to work with the Food Industry to increase people's access to essential micronutrients and to enhance their health and well-being. The main program includes the following:

- The fortification of commonly consumed foods.
- Dietary supplementation.
- Support to public health initiatives.
- Promotion of nutritious, locally available foods.

There is now significant global attention to micronutrient malnutrition. However the Niger Delta in particular and Africa generally is lagging behind the rest of the world in the campaign to iodize edible salt and the provision of high-dose vitamin A capsules among children lesser than five years of age.

Food fortification is gaining acceptance as a suitable strategy for protecting large populations against micronutrient deficiencies. Most significantly, the knowledge that micronutrients are essential requirements for health and well being is becoming widely accepted in Nigeria and the African continent as a whole.

Companies and other stakeholders can play an important role in implementing solutions starting with the Niger Delta region and spreading out to the rest of the country and sub-region. The can collaborate with International Development Agencies, Foundations, and Corporate Grant-Makers etc to facilitate the program by providing the necessary resources and tools to initiate action utilizing the skills of professionals.

Initially the stakeholders can plan to focus on three key programs in the Niger Delta:

- The reduction of iodine deficiency disorders through the iodination of salt consumed in the region.

- Reduction of vitamin A deficiency and its consequences including blindness.

- Reduction of iron deficiency anemia among women by one-third of 2000 levels.

This can only be achieved through a combination advocacy (including public enlightenment), technical, financial and operational support.

The mission is for professionals to work with the Private and Public Sectors and International Agencies to initiate, stimulate and implement appropriate actions to eliminate micronutrient malnutrition in the Niger Delta in particular and Nigeria in general.

It can achieve this objective by the introduction and expansion of food fortification and dietary supplementation programs in areas of greatest need in the Niger Delta.

It can work with relevant organizations to advance the ability to address iron deficiency anemia, and assist in alleviating the burden of micronutrient malnutrition as part of its integrated development agenda for the eradication of poverty.

The trial period for this program could be twelve months; with impact assessment reports submitted every three (3) months i.e. four (4) reports for the 12-month period

TWENTY-SIX

Strengthening Governance

It is generally accepted that the best path to sustainable development is through democracy, respect for human rights, peace and good governance.

Governance is the use of political, economic and administrative authority and resources to manage the affairs of society. This implies a system of values and institutions by which society manages its affairs through interactions within and between the state, civil society and the private sector.

Democratic forms of political governance are generally credited for good governance or amore efficient and effective management of public affairs in the interest of the people. Democratic governance thus implies the provision of social opportunities likely to increase the capacity of individuals to better their own lives. For ordinary people this means enhancing their capacity to overcome poverty, unemployment and social exclusion, and take advantage of the economic environment to improve their standard of living.

For our leaders, good governance can only be achieved through respect for the global standards of democracy, the core components of which include political pluralism, allowing for the existence of several political parties and workers' unions and fair, open and democratic elections periodically organized to enable people to choose their leaders freely. In order to strengthen political governance and build capacity to meet these commitments, the leadership must undertake a process of targeted capacity building initiatives by way of institutional reforms that should focus on:

- Administrative and civil services
- Strengthening legislative oversight
- Promoting participatory decision-making
- Adopting effective measures to combat corruption and embezzlement
- Undertaking judicial reforms

Experience has shown that there are close linkages between corporate governance mechanisms and public sector governance. Failures and lack of accountability in the private and public sectors often go hand in hand and can result in widespread corruption.

Controlling corruption requires matching the role of government with its capability, adopting sound legal and regulatory frameworks for private sector activity, establishing a strong and motivated civil service, setting up sound budgeting and financial management systems, putting effective watchdog institutions in place and providing capacity to detect, investigate, and prosecute fraud and corruption when and wherever it occurs. Assessing the costs of corruption and building a consensus for concerted action in the Niger Delta will contribute in no small measure to saving resources meant for the prosperity of its people

T W E N T Y - S E V E N

Epilogue

Current and future political and business leaders of the Niger Delta leaders must bear in mind the following:

* There should be recognition of the need for close public-private partnerships in formulating catalytic strategies for economic growth in Niger Delta in particular and Nigeria in general. Such collaborations are especially important in Nigeria both because investors have often lacked confidence in the overall business environment and private sector initiative, energy and finance are critical to positive growth trajectories.

* To achieve higher growth rate in the region, foreign investors perceptions of relatively high risks in the region need to be addressed and overcome. Consequently the approach to transformation must reduce perceptions of risk, if it is to support the attainment of higher economic growth rates. Peace and security in the region is one way of addressing the high-risk perception in the Niger Delta.

* Transformation of the Niger Delta and investment led growth requires a package of specific measures to provide certainty and clarity in the operating environment, to promote partnerships in business participation and to provide confidence in the future value of investments.

* The above can be achieved, respectively, through a transformation score card, applicable across all sectors, the introduction of investment incentives linked to transformation performance (specifically differential rates of corporate tax) and

a comprehensive plan to promote alternative available sources of equity funding to support SMEs in the region.

* For promoting private investment, there must be commitment to actively addressing institutional impediments to private sector investment in the region.

* Also needed is the pro-active promotion and participation in identified feasible projects, thereby insuring early successes that will in turn counter the current perceptions of risk and lead to an escalation of private investment.

* There can be no doubt that the quality and quantity of basic technology has a fundamental effect on the cost of goods and services in any modern economy. The Niger Delta cannot be different. Technological progress, and (it should follow) increasing integration, is in some ways natural and a self-fuelling process, depending chiefly on human ingenuity and ambition. It would be hard to call a halt to innovation. But it is easier to block the effects of technological progress on economic integration, because integration also requires economic freedom. ˙

* Economic opportunities in the Niger Delta would be far greater, and poverty therefore vastly reduced by reducing barriers to trade that restricts economic freedom.

* There should be a commitment by governments in the Niger Delta to increasing educational opportunities for creating the kind of environment that will allow its people to build the region into the economic force it should be. The success of countries as diverse as India, Malaysia, Ireland, Mexico, Taiwan and Poland demonstrates how countries can shape national strategies to capture the forces of globalisation and turn them to their advantage. This must include policies to develop or adopt technological innovations so that the gains are made from technologies essential to national competitiveness.

* Technology can be the main driver for rapid and sustainable prosperity in the Niger Delta. The advances achieved in computing and telecommunications in the West offer enormous, indeed unprecedented, scope for raising living standards in the Niger Delta in particular and Nigeria in general. New technologies

promise not just big improvements in local efficiency, but also the further and potentially bigger gains. These gains are not just profits for western and local corporations, but productive employment and higher incomes for the poor.

* Funding available from Multilateral and Bilateral agencies for would be limited, thus highlighting the need for private sector participation.

T W E N T Y - E I G H T

Bibliography

- Niger Delta Regional Development Master-plan (2006-2020).

- Niger Delta Human Development Report: UNDP, 2006.

- Nigerian National Petroleum Corporation (NNPC) Presentation at Abuja Petroleum Roundtable, March, 2007.

- OPEC: Revenue Fact Sheet, June 2003.

- Enterprise Solutions to Poverty, Shell Foundation Report, March 2005.

- Sustainable Investment in Africa: Pipedream or Possibility Shell Foundation Report/Forum for the Future, 2005.

- *People and the Environment Annual Report 2001* (Lagos: SPDC, 2002), p. 6.

- Royal Dutch/Shell Group of Companies, *Statement of General Business Principles*, 1997.

- Shell, *Profits and Principles-does there have to be a choice?* (London and the Hague: Shell International, 1998).

- *People and the Environment Annual Report 2001*, pp. 10-11.

- Dave Clark, "US envoy in Nigeria as America looks to African oil," AFP, July 25, 2002.

- Jim Lobe, "US Lawmakers, Israelis, push for more W African oil," *InterPress Service*, July 26, 2002; Carl Mortished, "US presses Africa to turn on the tap of crude oil," *Times* (London),

July 29, 2002. Nigeria is already producing well over its nominal OPEC quota of 1.7 million bpd.

- "Niger Delta Environmental Training Programme," Brief on Programme, FCO, 2001.

- Fukuda-Parr, S., and A. K. Shiva Kumar. (2003), *Readings in Human Development*, Oxford University Press.

- Gary, I., and T. Karl (2003), 'Bottom of the Barrel: Africa's Oil Boom and the Poor', New York: Catholic Relief Services.

- Ghai, D. (1988), 'Participatory Development: Some Perspectives from Grassroots Experiences', Discussion Paper 5, UN Research Institute for Social Development, Geneva.457

- Imobighe, T. M. (2004), 'Conflict in the Niger Delta: A Unique Case or a Model for Future Conflicts in Other Oil Producing Countries?' in Rudolf Traub-Merz and Douglas *Nigeria-European Union Country Support Strategy and Indicative Programme for the period 2001-2007* (Brussels: European Union, 2001).

- African Network on Environmental and Economic Justice (2004), 'Oil of Poverty in Niger Delta'.

- Aigbokhan, B. E. (1998), 'Poverty, Growth and Inequality in Nigeria: A Case Study', a report from the African Economic Research Consortium, Nairobi.

- Azu (2005), columnist, *The Punch Newspaper*.

- Bass, H. H. (1992), 'Development and Popular Democracy as Productive Factors or Constructive Elements of a New Development Vision', in H. H. Bass et al.

- *Industrialization Based On Agricultural Development: African Development Perspectives Yearbook, 1990/91*, Hamburg: Lit Verlag.

- Burke, Stan (1996), *People First: A Guide to Self-Reliant, Participatory Rural Development*, London: Zed Books Ltd.

- Central Bank of Nigeria (1999), *Annual Report and Statement of Accounts*, Abuja.

- ———— (2005), 'Baseline Study on Small-Medium Industries in Nigeria', Abuja.

- Chinsman, B. (1995), 'Putting People First', a statement to the Journalist Encounter at the World Summit for Social Development.

- Development Policy Centre (2001), *Socio-Economic Mapping of the Niger Delta*, a report prepared for Shell Petroleum Development Company, Ibadan.

- ERML (1997), *Environmental and Socio-economic Characteristics of the Niger Delta*.

- Federal Office of Statistics (1999), *Poverty Profile of Nigeria, 1980-96*, Lagos.

- ———— (2004), *Living Standards Survey of Nigeria 2003/2004*, a draft report, Abuja.

- Yates eds, *Oil Policy in the Gulf of Guinea: Security & Conflict, Economic Growth, Social Development*, Friedrich Ebert Stiftung.

- International Labour Organization (1977), *Meeting Basic Needs: Strategies for Eradicating Mass Poverty and Unemployment*, Geneva.

- Jonathan, G. E. (2004), *Niger Delta: Challenge of Sustainable Development*, Calabar: Nigerian Union of Journalists (Cross River State Chapter).

- Mosunmolu Limited (1998), 'Flooding study of the Niger Delta'.

- Niger Delta Environmental Survey (NDES) (2000), *Niger Delta Development Priorities and Action Plan*, phase II report, vol. 2.

- Oladipo, E. O. (1996), 'Community Participation in Development', a paper presented at the 'Policy-level Seminar on

Accountability and Transparency', UNDP/National Planning Commission, Akure, 11-13 March.

- Matthew Jones, "IFC postpones funding for Niger Delta," *Financial Times* (London), June 14, 2001; "IFC and Nigeria," IFC, August 2001. The IFC also approved a $2.5 million loan to construct a new hotel in Port Harcourt.

- "IFC Partners with Nigerian Entrepreneurship Program to Promote Small Business Development in Niger Delta," IFC press release, March 14, 2002.

- "IFC Partners with Nigerian Entrepreneurship Program to Promote Small Business Development in Niger Delta," IFC press release, March 14, 2002.

- Gary G. Moser, IMF resident representative in Nigeria, "The IMF and Nigeria, an Enduring Relationship," *This Day*, April 15, 2002.

- IMF, "Nigeria-Concluding Statement," March 6, 2002.

- Warner, M. and Sullivan, R. (2004) Eds Putting Partnerships to Work: Strategic Alliances for Development between Government, the Private Sector and Civil Society London: Greenleaf Publishing.

- Warner, M. (2003) The New Broker, Brokering Partnerships for Development London: Overseas Development Institute.

- Francis, P., J. A. Akinwumi, P. Ngin, S. A. Nkom, J. A. Olomajaya, F. Okunmadewa and D. J. Shehu (1996), *State, Community and Local Development in Nigeria*, Washington, D.C.: World Bank.

- Central Bank of Nigeria (1999), *Annual Report and Statement of Account 1999*, Abuja.458

- Central Bank of Nigeria (2003), *Annual Report and Statement of Account 2003*, Abuja.

- Federal Environmental Protection Agency (1991), 'National Environmental Protection (Effluent Limitation) Regulations', Lagos.

- Federal Office of Statistics (1999), *Poverty Profile of Nigeria*, Abuja.

- Federal Office of Statistics 2004. "Report of Nigeria Living Standard Survey 2003/2004, p.78, 81

- Government of Saudi Arabia, Ministry of Economy and Planning (2000), 'Human Development Index for the Kingdom of Saudi Arabia'.

- Jahan, Selim (2003), 'Unequal Human Impacts of Environmental Damage', in S. Fukuda-Parr and A. K. Shiva Kumar eds. (2003), *Readings in Human Development*, Oxford University Press, pp. 336-337.

- National Bureau of Statistics (2004), *Poverty Profile of Nigeria*, Abuja Associate gas utilization: Collaborative action –

- Scope in Nigeria and International best practice, Bent Svensson, World Bank, November, 9 2004

- Sen, Amartya (1989), 'Development as Capacity Expansion', in S. Fukuda-Parr and A.

- K. Shiva Kumar eds. (2003), *Readings in Human Development*, Oxford University Press, pp. 3-16.

- ul Haq, Mabbub (1995), 'The Human Development Paradigm', in S. Fukuda-Parr and A.

- K. Shiva Kumar eds. (2003), *Readings in Human Development*, Oxford University Press, pp. 17-34.

- OECD (1995), *Participatory Development and Good Governance*, Paris.

- Powell, I. (1995), *Wildlife Study Report for Shell Petroleum Development Company (SPDC) of Nigeria*.

- United Nations Development Programme (UNDP) (2001), *Human Development Report 2001*, New York, Oxford University Press.

- Gary, I., and T. Karl (2003), 'Bottom of the barrel: Africa's oil boom and the poor', New York: Catholic Relief Service.

- *The Guardian* (1985), 'Oil spillage victims reject compensation', 27 May, p. 3.

- ——— (2004), 'Wabara decries Shell's failure to pay $1.5b compensation', 17 November, p. 3.

- Hewawasam, Indu, et al. (2002), *Poverty-Environment Linkages in Nigeria*, World Bank.

- Jae-Young, Ko, and John W. Day (2004), 'Wetlands: Impacts of Energy Development in the Mississippi', *Encyclopedia of Energy*, volume 6.

- Kanoh, T., M. Fukuda, E. Itayemi, T. Kinouchi, K. Nishifuyi and Y. Ohnishi (1990), 'Nitro-reaction in Mice Injected with Pyrene During Exposure to Nigeria Dioxide', *Mutat Res.* 245, pp. 1-4.

- National Bureau of Statistics (2005), *The Nigerian Statistical Fact Sheets on Economic and Social Development*, Abuja.

- NDDC (2004), *Niger Delta Regional Master Plan Final Report: Waste Management Sector.*

- NDES (2000), *Niger Delta Development Priorities and Action Plan*, phase II report, volume 2.462

- NDHS (2003), *Nigeria Demographic and Health Survey, 2003/2004*, National Population Commission, ORC Macro and the US Agency for International Development.

- Nigeria National Petroleum Corporation and Academic Associates Peace Works (2004), 'Report of the Nigeria Delta Youths Stakeholders Workshop Held in Port-Harcourt, 15-17 April', Abuja.

- National Bureau of Statistics/Federal Office of Statistics (2005), 'Report of Nigeria Living Standard Survey 2003/2005', Abuja.

- Ololesusi, Femi, V. A. Adeyeye and Niyi Gbadegeshin (1993), 'The Impact of the River Bank Erosion Control Strategies on Agriculture in River State of Nigeria', a research report submitted to the Social Science Council, New York.

- Phil Anozia Consultants (2002), 'Environmental Impact Valuation Report on Oil Spillage Along the SPDC Trans-Niger Pipeline at Ogbodo, Ikwere, LGA', submitted to the Rivers State Ministry of Environment and Natural Resources, Port Harcourt.

- UNDP (2004), *Human Development Report 2004: Cultural Liberty in Today's Diverse World*, New York: Oxford University Press.

- ——— (2005), *Human Development Report 2005: International Cooperation at a Crossroads: Aid, Trade and Security in an Unequal World*, New York: Oxford University Press.

- Ali-Akpajiak, S.C.A., and T. Pyke (2003), *Measuring Poverty in Nigeria*, Oxford: Oxfam Publishing.

- Department for International Development (DFID), EC, UNDP and World Bank (2002), *Linking Poverty Reduction and Environmental Management: Policy Challenges and Opportunities*.460

- Eka, O. U., and I. R. Udotong (2003), 'A Case Study of Effects of Incessant Oil Spills from Mobil Producing Nigeria Unlimited on Human Health in Akwa Ibom State', in E. N. Aina, O. B. Ekop and V. I. Attah eds. *Environmental Pollution and Management in the Tropics*, Enugu: SNAAP Press Limited, pp. 204-229.

- Ekpo, Udeme (2004), *The Niger Delta and Oil Politics*, Lagos: International Energy Communications Ltd.

- Fajemirokun, Bola (1999), 'Review of Compensation Litigation in the Niger Delta: A Consultancy Report', Lagos: Practice Development & Consultancy Co. Ltd.

- Federal Republic of Nigeria (1996), *Nigerian Oil Industry Statistical Bulletin 1996*, Lagos: Department of Petroleum Resources. (2001), *Nigeria Oil Industry Statistical Bulletin 2001*, Lagos: Department of Petroleum Resources.

- Fukuda-Parr, Sakiko, and A.K. Shiva Kumar eds. (2003), *Readings in Human Development*, Oxford University Press.461

- Ohimain, Elijah (2004), 'Environmental Impacts of Dredging in the Niger Delta', *Terra et Aqua*, 97, December.

- Olawoye, J. J., O. Oladeji, O. B. Oyesola, O. Taiwo and A. Olomola (2003), 'Local Empowerment and Environmental Management Project (LEEMP): Social Analysis Report for Bayelsa State', submitted to Geomatics Nig. Ltd.

- Udotong, I. R. (2001), 'Environmental Monitoring and Effects of Petroleum Production Effluent on Some Biota of the Lower Qua Iboe River Estuary', Ph.D. thesis, Rivers State University of Science and Technology, Port Harcourt.

- Worika, Ibia Lucky (2002), *Environmental Law and Policy of Petroleum Development*, Port-Harcourt: Anpez Centre for Environment and Development.

- World Bank Environment Group, Africa region. (2001), *Hunting of Wildlife in Tropical Forests: Implications for Biodiversity and Forest Peoples Environment*, Department Papers, Biodiversity Series, Impact Studies, Paper No. 76, 42, Washington, D.C.: World Bank.

- Alubo O. A., Zwandor T. Jolayemi and E. Omodu (2002), 'Acceptance and Sensitization of PLWA in Nigeria', *AIDS Care*, 14(1), pp. 117-126.

- (2004), 'Gender Issues in HIV/AIDS Prevention and Control', a paper presented at the Strategic Skills Development Programme for NGOs, Social Sciences and Reproductive Health Research Network, Ibadan, 9-15 May.

- Ojo, O.J.B. (2002), *The Niger Delta: Managing Resources and Conflicts*, Ibadan: Development Policy Centre.474

- Schecter, Danny (2001), 'How Should Media Handle Conflict?', www.mediachannel.org.

- Shell Petroleum Development (2002), *Annual Report.*

- Chokor, B. A. (2000), 'Appraising the Structural Aspects of the Crisis of Community Development and Environmental Degradation in the Niger Delta', in A. Osuntokun ed., *Environmental Problems of the Niger Delta*, Lagos: Friedrich Ebert Foundation, pp. 62-78.

- Federal Office of Statistics/International Labour Organization (2001), 'Labour force sample survey December, 2000', Federal Office of Statistics *Statistical News*, number 322, June.

- Hassan, C., J. Olawoye and K. Nnadozie (2003), 'Impact of International Trade and Activities of Multi-national Corporations on the Environment and Sustainable Livelihoods of Rural Women in Akwa Ibom State, Niger Delta', Nigeria *NEST Research Brief*, number 5.

- Okali, D., E. Okpara and J. Olawoye (2001), *Rural-Urban Interactions and Livelihood Strategies: The Case of Aba and Its Region, Southeastern Nigeria*, London: International Institute for Environment and Development.

- Olawoye, J. (2002), *Gender and Rural Transport Initiative (GRTI): Analytical Study of Gender Specific Problems affecting Nigerian Women in Rural Travel and Transportation and the Identification of Pragmatic Solutions to the Problems*, a report for the Department of Rural Development, Federal Ministry of Agriculture and Rural Development, Abuja.

- Udeme, Ekpo (2003), *The Niger Delta and Oil Politics*, Lagos: Quadro Impressions Limited.

213

- Bureau of Economic and Business Affairs (2001), '2000 Country Reports on Economic Policy and Trade Practices: United Arab Emirates'.

- Centre for International Private Enterprise (2001), 'Mexico: Fighting Causes of Past Crises', www.cipe.org/publications/fs/articles.

- TDC Trade (2006), 'Market Profiles: United Arab Emirates (UAE)'.

- US Agency for International Development (2003), 'USAID: Latin American Caribbean Overview', www.usaid.gov/policy/budget.476

- West African Institute for Financial and Economic Management (2004), *Social Welfare Sector Study Final Report on the Niger Delta Region*, submitted to coordinating consultants (GTZ/Wilbahi).

WIKIPEDIA REFERENCES

- U.S. Energy Information Administration (U.S. EIA), 'Nigeria Country Analysis Brief,' December 1997.

- Environmental Resources Managers Ltd, Niger Delta Environmental Survey Final Report Phase I; Volume Environmental and Socio-Economic Characteristics (Lagos: Niger Delta Environmental Survey, September 1997)

- Nigeria: The Political Economy of Oil.ISBN 0-19-730014-6. (Khan, Ahmad)

- Nigerian Crude and Gas Industry (NigeriaBusinessInfo.com)

- Natural Gas (Online Nigeria Portal)

- Where Vultures Feast. (Okonta and Douglas, 2001).

- The Price of Oil: Corporate Responsibility and Human Rights Violations in Nigeria's Oil Producing Communities (Human Rights Watch, 1999)

- The Open Sore of a Continent. Soyinka, Wole.

- [Quoted in Greenpeace International's Shell Shocked, 11] (Greenpeace)

- Impacts of Oil spills along the Nigerian coast (The Association for Environmental Health and Sciences)

- Shell International Petroleum Company, Developments in Nigeria (London: March 1995)

- Nigeria to start rehabilitation of oil joint venture facilities (NigeriaBusinessInfo.com)